ALIEN EXPERIENCES
25 Cases of Close Encounter
Never Before Revealed

Barbara Lamb, MS, MFT, CHT

Nadine Lalich

HB Publishing, LLC
P. O. Box 65922
Albuquerque, NM 87193
(714) 625-6337

www.HBPublishing.net
www.AlienExperiences.com

First published in 2008 by Trafford Publishing
Reprinted 2017and 2020 by HB Publishing, LLC
(With minor revisions for enhanced clarity.)

Printed in the United States of America

ISBN: 978-0-9711776-4-2

READERS' COMMENTS

*A gold mine of information. **Nicholas Moore, March 23, 2020***
This is a remarkable book on ET abduction. Many wonderful pieces of information that will give even seasoned researchers new information they have not seen elsewhere.

*Very important for humanity! **Tom Lorton, February 19, 2020***
These events are defining the future of our planet and all fellow earthlings. Many are sketchy, hard to define. The most detailed accounts I find are of Nadine's. She remained more lucid than most in taking in the events that she was part of. The fact that she held sessions and co-wrote this book with Barbara is even more compelling to read it. Barbara has a good amount of experience working with experiencers. Both are brave individuals treading on what many don't want to consider a possibility and ostracize the messengers.

*This was a good read. **Denise Campbell, January 8, 2020***
I found this book very interesting. It would be a good read for anyone interested in UFOs.

*It was a great read. **William, April 25, 2017***
This was a great read, well written by Barbara Lamb and Nadine Lalich. I have passed it on to others to enjoy.

*One of the best! **Donna Patrick, February 16, 2016***
This book is amazing!! I would give it ten stars. ...Barbara and Nadine are masterful writers who captured their thoughts, emotions, and social implications. I have over 300 books by the best new age authors - Delores Cannon, Dee Wallace, Kyron, Lieu, Ruth Montgomery and I would put this book right at the top! Thank you, Barbara and Nadine.

*Things that go bump in the night. **Kenneth Hoffman, April 29, 2015***
I like this book, but you will only like this book if you believe in alien abductions or are trying to piece past life experiences with those of abductees. The authors do a good job of giving 25 accounts of alien contact. Specific descriptions of alien surroundings, procedures, and alien types will help if you're trying to jog your own memory. The large segment of differing abduction experiences makes you wonder how many different forms of life out there are observing us. Well-written, and possibly useful if you are a believer.

What an eye-opener! Mary Ann, December 6, 2014

Wow, the heartfelt stories into other realities are certainly mind-blowing but we had better get used to it because it is our unknown reality. Mind gripping reading!

Fantastic accounts of alien abductions. Lynn G., August 3, 2014

I have been interested in alien abduction stories since seeing a television program about abductions. I've had some strange experiences in my life that I can't explain and this book has urged me to have some regression therapy.... This book is a must-read. It is interesting and informative. The book is well written and a quick read. I'm sorry it ended .... Please read this book. It's simply amazing.

Very interesting. Jane Park, March 28, 2014

Anxious for new books from these authors. These case histories are very thought-provoking and parallel many accounts I have read in many other books on abduction.

So interesting. Jen A., January 10, 2014

This is a gem for anyone that likes to read about ETs. I work in a library and have gone through all of the books on this subject, but noticed the same stories in each of the books. Most, if not all of these stories in this book were new to me and incredibly interesting. I even read it twice! Love this book. A must for a UFO/ET buff!

Up-to-date Look at Abductions. Neil Finlay, July 19, 2013

Barbara Lamb has great experience from a regression therapist's point of view, having collected information from hundreds of experiences. This knowledge is combined with actual abduction cases as told by the people themselves.

CONTENTS

ACKNOWLEDGEMENTS

Barbara Lamb

I gratefully acknowledge Nadine Lalich for initiating our collaboration on this book, and for spending three years in organizing the voluminous amount of information collected from my eighteen years of regression work with people who have experienced extraterrestrial encounters. I highly respect her skills and intelligence, and she brings logic and discipline to this entire work. I also commend her persistence in exploring her own unusual experiences. Serendipitously, I found her to be a delightful person to know.

I acknowledge all the experiencers of extraterrestrial encounters that have come to me for regression work to understand the mysterious events that have happened to them. I salute them for their courage in sharing the recollections of their strange experiences and in exploring the material locked in their subconscious minds.

I appreciate the late Hazel Denning, Ph.D. for encouraging me in 1984 to take the years of training I have in Past Life Regression Therapy work through the Association for Past Life Research and Therapies. The skills that I developed in past life regression work enabled me to expand my expertise to include the use of hypnotic regressions with extraterrestrial experiencers. This fine organization has also provided the opportunity for me to teach training seminars to existing regression therapists for work with people experiencing these anomalous encounters.

My enthusiastic thanks go to R. Leo Sprinkle, Ph.D. for producing the Rocky Mountain Conference on UFO Investigations for eighteen years for experiencers of extraterrestrial contact, and to Bob and Teri Brown for producing the International UFO Congresses for eighteen years. Through these conferences, I became convinced that the UFO phenomenon and the abduction encounter phenomenon are real. During these conferences, I have also initiated and conducted large support groups for experiencers, and have been able to personally speak with hundreds of these abductees, many of whom I have had the privilege to regress to the details of their encounters.

My eternal appreciation goes to the late John E. Mack, M.D., who since 1993 encouraged and respected me in my regression therapy work with experiencers of extraterrestrial encounters. We validated each other's findings of highly spiritual and transformational experiences which some of our clients were having. We were collaborating on bringing forth the reality of Reptilian beings' interactions with human beings when he suddenly met his death. During our ten days of visiting crop circles together in England, he substantiated and enlarged my perspective of a multidimensional, paranormal cosmos and other-dimensional beings.

Although this subject is not directly in their fields of interest, I enjoy the encouragement of my husband, Warren, and my adult children, Christopher, Jennifer and Erica, for my writing a book on such an extraordinary subject. Their support of my interest, enthusiasm, and commitment to this material is gratifying to me.

Nadine Lalich

Originally, when I contacted Barbara Lamb in January 2005, little did I know that our meeting would be a turning point in my life. I very much appreciated her ability to listen without defining or imposing an agenda on my experience. In our early conversations, her thoughtful and supportive manner in discussing alien abduction allowed me to relax and more easily consider a subject that normally left me cold. The following year when I approached her with the suggestion that we collaborate in writing *Alien Experiences*, I was very pleased to receive her enthusiastic response. Working with her on the project was thoroughly enjoyable, and a wonderful opportunity for me to bring my skills and experience to a fascinating arena.

I would like to thank Pamela Serwatowski, my close friend for thirty-five years for her support and encouragement throughout the process of writing this book. I would also like to express my gratitude to the many abductees who, regardless of the stigma attached to this subject, have fearlessly come forward to share their unusual experiences.

Finally, I would like to thank those people who have dedicated themselves to investigating this phenomenon, not only to report but also to find the piece of evidence that will one day substantiate and bring credibility to those people experiencing contact.

FORWARD

Welcome, reader, to the world and work of Barbara Lamb and Nadine Lalich. In *Alien Experiences*, the authors have provided extensive information regarding the incidence and significance of alien encounters. I am pleased to offer this Foreword as an expression of my support for their work that I believe reflects competence, compassion, and courage. They have competently presented the cases in a logical and orderly format and provided enlightening background information on the subject in general. Their desire to assist those who are struggling with this phenomenon reflects their compassion, and the list of questions they have included offers a helpful, practical tool for those readers seeking to unravel their mysteries. The individual commentary by each of the authors also sheds additional light on the cases presented in the book and on the subject at large. In offering *Alien Experiences*, Barbara and Nadine have demonstrated courage by the very act of writing and publishing a book on the subject of abduction. If you are a skeptical reader, you might ask, "Are the authors engaged in scientific research or psychotherapeutic service?" In my opinion, the answer to both questions is "yes."

My viewpoint about the extraterrestrial presence can be summarized in the following story. A friend of mine, a science teacher, once asked me a question.

"Leo, why would intelligent beings from some other star system travel millions of miles to earth to extract semen from men and ova from women and do it over and over again?"

"Why?" I smiled and said to him, "I'll answer your question if you will answer mine. Why would an intelligent science teacher conduct the same laboratory experiments, over and over again?"

He laughed and said, "Oh, for the sake of the pupils!"

A pupil learns with the help of an instructor or teacher. A student knows how to learn on his own. Whether we are pupils in need of further ET encounters for additional learning, or whether we are students who can and are ready to learn on our own *Alien Experiences* offers more opportunity for expansion. What are we learning? We need to grow up and care for the earth and humanity and prepare for space/time travel and communication with many levels of consciousness. Are we abductees who are taken *from* the earth or, are we abductees who are taken *to* the stars? Are we pawns of alien beings, or partners and students of the extraterrestrial presence? Are we planetary persons or cosmic citizens?

Many UFO experiencers seem to go through various stages of initiation (e.g. the PACTS model: preparation, abduction, contact, training, and service). As an analogy, many military recruits fear the tongue-lashing from drill instructors and hate the probing of physicians during medical examinations. Yet later, they may take pride in working with other trainees and serving their nation. Many UFO experiencers also initially, fear and hate ET encounters, then, later, take pride in serving humanity and star nations. Whether you are a skeptic, a debunker, an experiencer, or a believer of the ET presence on earth, dear reader, you are alive during a crucial stage of human history, for the shift in consciousness is occurring. May your awareness be enhanced by the work of the authors, and may your service be encouraged by the work of our cosmic cousins.

-R. Leo Sprinkle, Ph.D.

INTRODUCTION

As human beings, we have initially been conditioned to live in a limited reality of our own creation, projecting onto the world the beliefs, and attitudes instilled upon us early in life by family, friends, and society-at-large. Much of this conditioning is fear-based and capable of creating psychological blindness that can shut out awareness and opportunity, and potentially force us into limited, rigid views of reality and what is possible. Even the brightest, most open-minded and curious of persons should periodically reexamine their beliefs and their concept of reality, especially those beliefs that stem from fear and prejudice. Having the courage to do so will not only expand the consciousness of the individual but also enhance the consciousness and evolution of humanity. As you read the stories in this book, we truly hope that you will suspend whatever preconceived judgments you may have about the extraterrestrial contact and simply allow yourself to ponder the information presented. In doing so, you may be surprised to find that under very different circumstances, amazing similarities appear throughout the abduction experiences of men and women.

This may be the first time you are considering the possibility of contact between humans and alien life forms or, you might be a confirmed skeptic. Perhaps, though, you may be someone who senses that something unusual has happened in your own life and you are wondering if your experiences could be related to alien abduction. If that is the case, the best way to explore a challenging new concept is to face it directly through communication and open investigation, thus eliminating much of the mystery, confusion, and fear. Personal power can be reclaimed by such an assertive approach, restoring emotional and mental balance.

In 1991, Barbara's psychotherapy and regression therapy practice took a surprising turn when individuals began to appear who were seeking assistance in solving the puzzle of what appeared to be interactions with intelligent, non-human life forms. As time went by, more and more individuals sought her assistance and, as a result, she became dedicated to exploring the phenomenon and in supporting experiencers.

Nadine's interest grew due to her strong investigative instincts and dedication to personal growth that would not allow her to ignore certain unusual events in her own life that defied conventional explanation. In presenting the information in this book, the combined diverse backgrounds, skills, and experience of Barbara and Nadine have created a unique and effective working partnership.

The accounts of extraterrestrial contact presented in *Alien Experiences* have been collected by Barbara Lamb under various circumstances over seventeen years. Her sources include colleagues, attendees at UFO conferences, UFO group participants, and hypnosis clientele. While maintaining the integrity of each story, all of the accounts herein have been presented anonymously, without reference to the actual name or location where such information was obtained. Pseudonyms have been used for ease of reference and, when required, permission to print was obtained. For the sake of simplicity, when referring to an abductee whose gender is unknown, we have referred to that individual as *he or him*.

Although we cannot scientifically prove that the accounts of alien abduction depicted in this book have truly happened, we have recorded the information *as it was presented* in as true a form as possible. We can also affirm the sincerity of the experiencers who we are convinced have experienced *some kind* of anomalous event.

1

HUMAN ABDUCTION
BY EXTRATERRESTRIALS

Many current and ancient sources indicate that interaction between extraterrestrials (often referred to as "ETs") and humans may have been occurring for thousands of years. In recent times, contacts and abductions have been reported since the late 1940s. According to some polls, it is a worldwide phenomenon involving as many as eight million people in the United States alone.

History of Abduction

People appear to be abducted by extraterrestrials for a variety of physical reasons that include medical examinations, testing of body fluids, insertion and replacement of implants, reproductive procedures, and possibly for sexual relations between humans and extraterrestrials. Other reasons include human interaction with hybrid offspring and testing of human emotional reactions.

Alien abduction usually begins early in life, sometimes even before birth. Often, the phenomenon will target an entire family for generations, and, over time, some or all of the family members will experience contact throughout their lives. As children, they respond to these contacts in various ways, ranging from terror to sometimes enjoying the contact with beings whom they consider friends. Children are taken from their beds at night or from wherever they are playing during the daytime and taken aboard a spacecraft. They are medically examined and sometimes given a tour of the craft, or introduced to hybrid children and encouraged to interact with them.

Occasionally, when children enter adolescence, a particularly traumatic procedure can begin. There seems to be an on-going interbreeding program where eggs or sperm are removed from abductees and mixed with alien reproductive substances so that hybridization can take place. After such a procedure, females are abducted again and implanted with the resulting hybrid fetus, often with the pregnancy being later confirmed by their doctor. After a few weeks pass, they are abducted again and the fetuses are removed and placed in containers of fluid, leaving the females baffled when they realize they are no longer pregnant. Months later, these same females might be taken aboard a craft and encouraged to interact with an unusual looking baby who may well be their child. These women are urged to hold and bond with the strange, frail babies with large, penetrating eyes and large heads who do not look entirely human. During the next several years, these women could experience this reproductive process several times, usually without much conscious awareness. If these things have occurred, the memories could come forth in hypnotic regressions.

Some abductees discover during regressions that they collaborate with the aliens during abductions. They may help convince other humans to go aboard ships, assist with medical procedures, or co-pilot crafts. Some people report that they are delegates to large council meetings comprised of aliens from many different planets, and others report traveling long distances in crafts while suspended in pods of special fluid. Some abductees come to like the aliens, and some believe that they have alien mates and hybrid children. Others describe spiritually transformational experiences with beings they considered very lovingly.

Methods of Abduction: Physical and Astral

Abductions appear to take place as physical or astral experiences while awake, via lucid dreaming or during deep meditation. In all cases, the abduction begins with the abductee being rendered paralyzed and amnesic, allowing for easier control by the abductors. Some experiencers have been told that the paralysis also helps to enable the particles of their bodies to separate enough to move through a solid wall or closed window. Because this part of the experience seems to be intrusive and non-optional, the abductee usually reacts negatively, even if the experience turns out to be a positive one.

During an astral or out-of-body experience, only the abductee's consciousness is taken and later returned to his body in the location where the abduction began. Astral abduction often includes the experiencer receiving instruction in a wide variety of subjects, including healing and psychic development, ecological and technological information, instruction on piloting crafts, and assisting surviving humans after earth disasters. Perhaps for their sense of security, these abductees construct in their minds a setting that seems appropriate for this kind of learning: a school, laboratory, library, temple of learning, or similar setting. Although they may be experiencing a non-physical location, they have a very conscious state of being. A physical abduction is often verified when the abductee subsequently finds strange markings on his body or other physical indications that his body was affected. An abductee may also experience combinations of physical and astral events that contain both positive and negative aspects. Starting as a terrifying experience, some events unfold in such a manner that the abductee ultimately develops a more positive emotional response to the experience. Frequently, abductees obtain meaningful, often life-changing information from the beings and the experience.

Varieties of Extraterrestrial Beings

According to reports from thousands of people worldwide, there are many different types of aliens and, over time, an abductee may encounter a variety of species, often during one abduction. They can be tall or short, with or without hair, insensitive or compassionate. The most common type of being described by abductees is the Grey who is three to four feet tall with grey skin and large, black, almond-shaped eyes, without any surrounding white area. His nose is barely discernable and he has only a small slit for a mouth. There is a small hole on each side of his head for hearing, but no earflaps. The head is hairless, the chin pointed and the body is thin and delicate looking. He moves stiffly, like a robot, and has been described by abductees as behaving in a very matter-of-fact manner as he attends to his tasks diligently and efficiently, apparently with no regard for the abductee's reactions. There are other Grey beings, of various heights, but all Greys possess large black, almond-shaped eyes. It is not uncommon for them to stare into an abductee's eyes from as close as an inch away from his face, feeling by the abductee as though his thoughts, memories, and entire personal history have been accessed. This mind scan is a startling, invasive, and frightening experience for most abductees.

Similar in stature, are the beings referred to as Little Whites, named for their chalky white skin. The Little Whites appear to be more aware and sensitive to humans, and less threatening in their behavior than the Greys. They have rounder eyes with a hint of white around the irises, and their eyes seem to convey some degree of feeling and responsiveness toward their human subjects. Both the Greys and the little Whites conduct the actual abductions and, sometimes, perform medical procedures while in the abductee's home or on a craft.

Another species known to be quite disturbing to many abductees is a large being that possesses many of the physical characteristics of a reptile. This being, with round, vibrant yellow-gold eyes with vertical pupils and rough alligator-like skin is often referred to as a Reptilian. Reptilian extraterrestrials often display an armor plate covering on their chests that may or may not be part of their natural body, and they sometimes wear long capes, robes, or banners across their chests with emblems or colored medals. Reptilians are approximately the size of a large adult human male with a muscular physique and an arrogant, aggressive nature. It has been reported that both male and female Reptilians, on occasion, engage in sexual relations with human members of the opposite sex.

Another commonly described alien species is a tall, thin insect type of being that looks like a Praying Mantis with huge bug-like eyes that wrap around the sides of his head, and extremely thin arms and legs. Species such as this type that look like insects are sometimes referred to as a Mantis or Insectoid. As repellant as they might at first appear, those who have encountered these species develop a fondness for them for their thoughtful and kind mannerisms, and some abductees even consider them unconditionally loving.

There are other aliens referred to as Hybrids who appear to have been genetically engineered by combining human and alien reproductive material. From a distance, they can appear quite human, but closer contact reveals that they are not. Their eyes are quite similar to a human but somewhat larger, and their demeanor and expression are detached and non-responsive. Allegedly, some Hybrids look human enough to be able to intermingle with humans on earth without detection.

There have been other types of beings reported, including a bronze-colored alien, approximately five-foot-eight or nine inches with large, faceted eyes. Several other types of White aliens reported include one with huge, pointed ears and very large feet and another who is extremely tall with wispy hair and white skin that is stretched tightly over a skull-like face. A type of being that has been reported by abductees and observers, in general, is the legendary Sasquatch or Big Foot being whose large body is covered with hair. There have also been reports of another insect type of species that has a short, stocky build and large, bulging bug-like eyes whose attitude and behavior appear so threatening that humans immediately recoil from them.

Finally, well-publicized types of beings referred to as Men in Black, have been known to contact UFO investigators or abductees in their homes or offices. These beings often confiscate material evidence related to alien activity, alien technology, or human governmental involvement with aliens and abductees. These Men in Black move with such robot-like manners that they may, in fact, actually be androids. These entities generally dress in black suits and hats, often fashioned after the 1940 era styles. They always cover their eyes with dark sunglasses and speak in monotone, canned voices, as if speaking requires a great deal of effort. They behave in a demanding and threatening manner when attempting to confiscate materials, sometimes threatening to appear in the person's sleep until the materials are relinquished. Because their energy tends to deplete quickly, the visitations are generally short-lived and, as they depart, they have been observed staggering and even disappearing into thin air after a short distance. They often arrive in large black sedans from the 1940s and 1950s eras, and these vehicles have been seen to vanish instantly, as well. It is not known if these beings are alien, Hybrid, or androids.

Some beings appear to be vibrating at a higher frequency, radiating light, and seem other-dimensional. These kinds of beings are nearly transparent and give the impression of being spiritually advanced and loving. Among this type is a huge, amoeba type of being with glowing blue and lavender colors, residing out in space. When a person interacts with one of these beings, he is overwhelmed with a sense of unconditional love.

In general, although there is rarely any apparent indication of gender, abductees do tend to identify extraterrestrials as male or female by their demeanor and by the vibrations they emit. Many alien species also appear to wear robes, form-fitting space suits, or some other type of apparel.

Alien Behavior

Many of the aliens described herein have been observed working together cooperatively during the abduction of humans. Certain species initiate the abduction, removing the abductee from his home or other location, and transferring him to a craft. Others perform medical procedures on the abductee while he is lying on an examining table, while another species performs a mind scan by visually penetrating the abductee's eyes very close to his face. Another being is often assigned to calm and comfort the abductee by touching various parts of the human body, such as his forehead, shoulder, arms, or wrist. A leader is often present who simply observes the examinations from the background. There have also been instances when aliens appear to conduct complicated healing procedures on the abductee, such as healing cancer, blood disorders, autoimmune diseases, or heart defects.

After a medical or sexual procedure has been completed, the abductee is often taken to another room and instructed regarding star maps, healing techniques, or is encouraged to learn alien symbols or writing. Often the abductee is shown devastating holographic scenes of the earth such as nuclear destruction, massive earthquakes, or other catastrophic views. During abductions, the extraterrestrials often stage provocative events, created physically, shown on a screen, or mentally projected into the abductee's mind to trigger an emotional response. Such staged scenarios might include seeing a child being struck by a car, having a deceased loved one or a current lover suddenly appear, or seeing a horrific action perpetrated upon a fellow human, with the human's emotional reactions closely observed and recorded by the extraterrestrials.

2

WORKING WITH
ABDUCTEES AND EXPERIENCERS

Those Who Are Chosen

Most abductees do not suffer from medical or psychiatric disorders; they are normal people to whom very unusual things have happened. Those abducted worldwide include men and women from all races, of any social status, education, or profession. Often, alien abduction appears to be a family phenomenon, with multiple family members being abducted over several generations. With the subject of abduction discussed more openly in the media, larger numbers of abductees have come forth to share their unusual experiences with family, friends, researchers, and therapists. Unfortunately, there remain countless other persons, including those from more prominent social or professional circles, who are reluctant or unwilling to speak out for fear of ridicule and loss of reputation.

Even after an abductee finds a person or a group with whom he can seriously share his experiences and they are validated, he may still need therapeutic help to incorporate the experiences into his life. Contact with extraterrestrials can cause various responses in people who experience it, ranging from wonder, awe, gratitude, and the feeling of having reunited with their true family. An abductee might also experience extreme anxiety and phobias and, in some instances, have such difficulty incorporating the experiences into his ordinary life that day-to-day functioning becomes difficult.

Indications of Extraterrestrial Contact

An abductee may consciously recall some or none of the elements of his abductions. Some experiencers, at first, only remember fragments from an abduction, believing that their conscious memory constitutes their total experience. Later, after being hypnotically regressed to the event in question, he can discover that there was much more to the experience that he had forgotten. Then there are those abductees who remember absolutely nothing for years until their subconscious mind is triggered by something such as a picture of an alien or a UFO. As a rule, amnesia is rendered upon the abductee by the aliens, and often reinforced by the abductee's own inherent denial and repression mechanisms.

There are many indications that a person may be an abductee, and therapists should be alert for these special clues. An abductee might experience on-going irrational anxiety, disorientation or confusion, and panic attacks. In the worst-case scenario, the person suffering from repressed memories of alien encounters may find work or school activities seriously disrupted. He may have developed an extreme fear of doctors or hospitals and medical instruments and procedures, sometimes so extreme that he avoids seeking medical treatment when required. Locations where light is diffused, such as airports or other large public spaces, may bring on panic attacks. He may also have relationship problems and difficulty trusting others, especially authority figures. Fear of reptiles or insects such as the Praying Mantis, or fear of looking into the eyes of animals such as owls and deer might develop.

Compulsive or addictive behaviors, as well as expanded psychic abilities, might be exhibited by someone with repressed memories of ET contact. Fears involving nighttime are strong clues, including a fear of the dark, being left alone or of going to bed and to sleep. The bedroom closet might become a focus of fear, as well as the fear of being watched or of driving alone at night. Someone conscious or unconscious about their abductions might have unusual sleeping habits, such as sleeping fully clothed or sleeping next to the wall. He might awaken each night at a certain time, or strive to stay awake until a particular time has passed when he feels safe from an alien encounter. He may suffer from insomnia or other sleep disorders, often experiencing the sensation of being slammed awake or dropped onto his bed from above. He may wake up in strange positions on the bed or in another location in his home and, possibly, even several miles away. At the beginning of an encounter, an abductee may have recollections of feeling paralyzed or of floating through the air, unable to call for help or wake up a sleeping partner.

After an abduction, physical clues may be present such as grass or dirt on his feet, wet hair, or he might find that the clothing he is wearing is on backward or inside out, or that it does not belong to him. Other physical body evidence to look for might be the appearance of small, scoop-shaped scars or triangular burn marks, infections in the navel area, thin cuts or pinprick marks, or bruises in the shape of a finger or handprint. There may be a small lump near one ear or in the forehead, and he may feel frequent vibrations, ringing or pressure in that area. Experiencers often have chronic sinusitis and nosebleeds, and such an abductee might sometimes awaken to find blood on his pillow. He might have back or neck problems, soreness in the genitals, or stiffness anywhere in his body, and pronounced sensitivity to certain lights and sounds. Electronic equipment may also turn off or malfunction when he is near. Women sometimes report that they became pregnant without having had sexual activity, only to find two or three months later that the fetus had mysteriously disappeared. A female abductee may also say that she discovered that one of a set of twins had vanished from her womb.

Helping Abductees with Regression Work

The field of therapeutic work with abductees is relatively new. The first case on record occurred in 1966 when psychiatrist Benjamin Simon hypnotically regressed Betty and Barney Hill to their disturbing abduction in New England. New to this experience, Dr. Simon concluded that the Hills must have been suffering a shared delusion. Unfortunately, since that time most mental health practitioners have failed to come to grips with the abduction phenomenon. Thus, clients experiencing abduction are usually misdiagnosed, given medication, or provided misdirected therapy that leaves them feeling more misunderstood and potentially in worse condition than before they sought help. Often they also refrain from sharing their experiences with family and friends to avoid possible ridicule and scorn, thereby increasing their anxiety even more.

There are no agreed-upon procedures for working with abductees, and few psychotherapists or hypnotherapists are open to the phenomenon or willing to take these people seriously. During an abduction, it appears that the extraterrestrials induce an altered state of consciousness in the human to mask his experience and more easily control him. Then, after the event, amnesia is induced in the abductee to reinforce memory loss. Because the subconscious mind records all events, the regression therapist possesses the necessary skills that can assist in uncovering the buried memories of abduction, which releases the toxic feelings that can be associated with such traumatic events. This helps the experiencer to integrate the phenomenon into his conscious life with positive results.

The Initial Interview

The initial contact with an abductee is important, whether in person or by letter or telephone. Often abductees are afraid to speak about their experiences and, although the therapist will not immediately know whether or not they have experienced abduction, it is crucial to be open-minded and to listen seriously and intently. A good option is for the therapist to send him a life history questionnaire and an information kit about regression work. The therapist can also ask him to write a brief description of anything that he believes may be related to an abduction experience, including dreams, sensing someone in his room, flashbacks, conscious UFO sightings, missing time, or anything else that seems relevant. It is also important to let him know that, should he choose to pursue the matter and later discover through regression that he has been abducted, his life and his notions of reality will be changed forever. If the therapist tells him this before beginning regression work, it will allow him the opportunity to consider whether he is ready and prepared for such a change in his life.

In the initial face-to-face interview, as with any other client, the therapist needs to determine whether any other kind of crisis or major life issue is going on, whether there is a history of abuse, and whether he has psychiatric symptoms or serious depression. If so, these issues must be dealt with first. In reviewing the client's history of possible abduction-related happenings, the therapist should look for where the emotional intensity is, such as important interruptions in the client's life, or for symptoms such as irrational anxieties, flashbacks or vivid dreams. The therapist should explore the client's fear of uncovering possible abduction material, and discuss what it is likely to mean in his life if he discovers that he has had the experiences. A client may begin to question his sanity, and it is necessary to reassure him that he need not. He may also need assistance in setting up a support system to help him feel stable and accepted as he goes through his regression work. He should also be acknowledged for his courage in seeking help.

During the first few sessions, the therapist must refrain from indicating whether he believes the client has experienced a true abduction. The client should draw his own conclusions about this, but he should be encouraged to refrain from doing so until after he has experienced at least one or two regression sessions. Asking the client to rate the reality of the regression experiences using a scale from 0 to 10, with zero representing the dream or fantasy state and ten indicating a sense of complete reality, may help him to make this decision. A few tens may be a good reason to continue exploring the possibility that the client is truly an abductee.

The frequency of abductions varies from person-to-person, but regression indicates that most abductees have been abducted more than once, often many times, starting in childhood. Intervals between abductions could be weeks, months, or years, yet some people may have had several abductions in one night. In the beginning, it may be particularly frightening, but over time, he may find some enlightening aspects to the contact, as well. He may also find that rather than just one species, he has been interacting with a variety of alien species. It can be helpful for an experiencer to have several regressions over time, to develop a more comprehensive picture of his overall experience of contact. In time, the fear tends to dissipate and a sense of curiosity develops for the experiencer who is now able to observe many other aspects of the alien-human encounter.

Encouraging an abductee to look for something positive in his experiences can help him to find more peace with the experiences and more balance in his life in general. Having less fear, greater awareness, and a more positive focus brings empowerment. The therapist can also suggest an experiencer to ask questions of the beings during abductions to gain greater clarity and understanding about what is happening and for what purpose.

It is probably wise to avoid weekly regressions with an abductee, as he needs time to assimilate and process the experiences he has recalled. In the interim, maintaining contact between the abductees and his therapist for encouragement and support is advisable, for the therapist may be the only person in his life who is receptive and understands his dilemma. Telephone conversations are a good way to communicate between regressions without placing undue stress on the client.

Most abductees go through a roller coaster ride of responses as regressions proceed, moving through denial, fear, anger, and depression until, gradually, they accept that the phenomenon is occurring in their lives. As the old concept of reality falls away, the growing awareness may give way to curiosity. It is important for the therapist to accept the whole continuum and to remain a stable container.

Hypnotic Regression Therapy

Hypnotic regression is an effective tool for assisting a person in retrieving and resolving buried memories. It is a safe, comfortable method for going back in time and reliving the details of a former experience that has not been remembered consciously or has been remembered only in brief fragments. Hypnosis brings the individual to a deep state of relaxation, fully aware and mentally focused, but free from external distractions. In this state, with the guidance of a competent hypnotherapist, he can direct his attention inward and access his subconscious mind where a detailed record exists of all the events of his life. During the regression, the person under hypnosis revisits an earlier place and time, as if it is happening in the present moment. The experience, with all of the details, emerges from the person's subconscious mind, moment-by-moment.

Once it has been determined that the cause of a client's distress is likely the result of having been in contact with extraterrestrial beings, regression therapy is the tool that can unlock the buried memories of an encounter, providing great insight and tremendous relief for the abductee. Along with integration work, it can help him assimilate and make sense of his bizarre experiences, thus improve his daily functioning, regardless of whether or not the experiences continue to occur.

As always before beginning any regression, the client should be prepared for the work ahead. Providing an explanation of hypnosis and accessing the client's primary focus, i.e., visual, auditory, kinesthetic, or intuitive knowing is helpful. The therapist can instill confidence by reassuring the client that he will be available to support every step of the way. It can also be suggested to the client that he can call on any higher guides he would like to for protection. He should also understand that it is not necessary to relive all aspects of an abduction experience during one session unless he desires to and that he can end the session whenever he chooses.

When beginning, it is best for the therapist to allow the client to choose the subject of the first regression, such as a dream, a fragment of a memory, or another meaningful event. The session can also be open-ended, in which case the therapist would direct the client's psyche to take him to the experience that it is most appropriate to examine at that time.

Whatever material comes forward should be explored as thoroughly as possible, without the therapist using leading questions or suggestions. Traumatic material may or may not appear during the first few sessions, and fearful clients should not be forced to remember. Moving slowly is the rule, letting the client take the lead. Abductees should be reassured that the memories will eventually surface, bringing clarity. For the first session or two, it is often necessary to allow them to examine traumatic material with the feeling they are only observing it, and not reliving it. Freezing time is a good technique to use to allow the client to examine only one traumatic incident at a time.

Before ending the session, giving the direction to move ahead to the conclusion of an event is helpful because it allows for closure, even if temporary. These clients will probably not remember all that emerges during a regression, and they should not be forced to recall it after the session as it will eventually come to the surface of his memory. They can also listen to a recorded audio tape or CD of the session when they wish to remember all of the details. Suggesting they do homework between visits, such as drawing or writing down memories in a journal can be a very useful tool. Flashbacks or triggered memories might also emerge that can be explored under subsequent hypnosis sessions.

Transformational Aspects of Contact

Some abductees consider their experiences less invasive or traumatic, and they more readily tolerate medical procedures that are performed. These people tend to look for the deeper meaning of their experiences, an attitude that may be part of their coping mechanism. These individuals also consider the alien-human interaction as ultimately benevolent, carried out by extraterrestrials for the benefit of our planet to stop its destruction. These beliefs bring a spiritual aspect to the abduction scenario causing the experiencers to accept the aliens as enlightened or spiritual beings. During abductions, they may experience a sense of awe because they believe they are returning to a greater realm of existence, one that ultimately heightens their appreciation for the natural world. For some, abduction experiences can also bring about a belief that consciousness is separate from the body. Such realization allows them to identify more with all living beings, and to be more comfortable with the cycle of birth and death. This also promotes less ego-attachment to present-day identity issues.

For an experiencer who feels only fear and negativity from their abductions, expanding their perspective to include a broader concept of life and the universe can open the door to learning and greater insight. It can also promote a spiritual awakening that offers some sense of well-being to those people who may continue to experience abduction throughout their lives. The most important goal of the therapist is to help relieve the suffering of the client and assist them to find a way to integrate the challenging phenomenon into their daily lives.

3

NADINE: A CASE STUDY

Background

In 2005, I was referred to a psychotherapist, Barbara Lamb, from someone I met at a meeting of the Mutual UFO Network ("MUFON") in Orange County, California. In 1991, I had experienced a dramatic and terrifying event while camping in Sedona, Arizona, and I was anxiously seeking assistance in unraveling a mystery that had traumatized me and left lingering effects over the years. I had been accompanied on that trip by my life-long friend, Pamela, who was rendered unconscious during the event, but who was still a witness to the fact that my personal belongings had mysteriously disappeared, i.e., *only* my belongings. Although I mentioned only a little of the experience to her at the time, she was quite aware of my distressed emotional state after the event which abruptly ended our trip.

Immediately after my experience, I was able to consciously recall many details of the event when I was physically removed from our vehicle and taken to a craft. That experience, along with many other unexplainable events that followed for more than a decade, was deeply disturbing to me as a conservative woman who at the time was quite unfamiliar with the phenomenon of abduction. For many reasons, including fear of ridicule, I chose to keep the experiences primarily to myself. Throughout the years as more bizarre experiences occurred and the burden became too great to bear alone, I would occasionally discuss my situation only with Pamela.

Overall, my greatest strength and comfort came from a strong spiritual foundation, and from turning to my journal where I recorded detailed accounts of my strange encounters with what seemed to be non-human entities. Having some artistic skills, I have been able to bolster my written recollections with sketches in my journal of what I saw and experienced. Over time, because most of the experiences take place in the middle of the night, I learned to keep my journal close at hand so that upon returning and/or awakening from an experience, I can quickly record as much information as possible while my memory is fresh.

As the years passed, although I continued to record events, I steadfastly refused to openly explore the notion of alien abduction, and kept looking for a more acceptable, traditional psychological or emotional condition as an explanation for the phenomenon. I was angry by the disruption in my life and uncomfortable seeing myself in the role of a victim; therefore, my records helped me to feel some sense of control over an otherwise powerless series of events.

It was after a particularly active period of bizarre occurrences, that I became determined to explore openly, to share my experience, and seek answers. There was a great relief in the fact that besides being well-acquainted with the subject of ET contact, Barbara was a seasoned psychotherapist who understood a variety of psychological and emotional human conditions. Even then, I clung to the hope that there would be other conditions that might be responsible for the disturbing occurrences.

As with so many abductees, my experiences involved years of conceivably ET abductions or contact, therefore, only a portion of these experiences have been presented in this chapter of *Alien Experiences*. These events were gathered from my *conscious* recollections over the years which I recorded in my journals, along with many sketches depicting details of the events. Although there is great controversy among researchers regarding the validity of conscious recollections versus those obtained through hypnotic regression, Barbara and I believe both methods of recall contain information and have valid applications in the search for truth. Excerpts from several hypnotic regressions conducted by Barbara Lamb, summaries of events, and journal entries are presented herein. Using hypnotic regression did prove to be an excellent tool that allowed me to focus more deeply upon several experiences that provided additional details from my subconscious mind. In each case, it is important to note that my initial impressions and recollections of those events recalled consciously were not contradicted when explored under hypnosis.

JULY 1968 INCIDENT

Journal Entry December 31, 2002

It was a strange night and I *came to* this morning at 6:00 A.M. to find myself sitting upright in my bed. I need to write about an old memory that I have dismissed for years as impossible and simply a figment of my imagination. I now believe there may be some truth to it.

When my boyfriend, Mike, and I were sixteen years old, we would often take long rides at night into remote areas. On one of these night time excursions, we saw what appeared to be a very large craft overhead in the sky, perhaps one hundred or more feet across with a saucer shape and a domed center. As we watched, amazed, it dropped lower until it was right above us where we had stopped on the dark road. Small lights were mounted around the center of the bowed underside of the craft. He and I had shared a very strong interest in watching the night sky and we were excited and shocked by this sight so we decided to follow it. When we shared the event with our parents later that evening, they did not believe us. Several years after our friendship had ended, Mike's sister told me that he had been hospitalized for psychiatric treatment for having seen strange beings that he said were *coming into his apartment through the walls,* and he claimed they told him to do strange things.

Excerpt from Hypnotic Regression February 12, 2005

B: So today, Nadine, we are going to do a regression and we are going back to that experience that you had in the summer of **1968** when you were sixteen years old and you were with your boyfriend, Mike, driving on a rural road. In that experience, you saw a large ship up in the air with a round dome on the top and you watched it until it came down near the ground. We are going to go back now to that experience and pick up all of the details that you consciously remember, as well as anything else that you were not consciously aware of. We are going back to the entire experience of that summer night in 1968 when you were sixteen years old, driving along the rural road with Mike, and seeing an unusual sight. To do that we're going into a nice state of deep relaxation, back to that summer of 1968 when you were driving with your boyfriend, Mike, who was a very special person to you at the time. On this rural road, both of you see a very large object in the sky that's coming down near the ground. It has a round dome on the top and you have decided to follow it. What else can you see about the object?

N: At first, it seems like its green, but maybe it's white light.

B: Where does that white light seem to be?

N: It's coming from underneath.

B: Does it do anything, or just stay right by this object?

N: At first, I thought it was short, coming out of little holes underneath it, but now it seems brighter.

B: Do you mean that it shines outward to some extent?

N: Yes. I also see tall weeds on the side of the road. It seems like we're slowing down.

B: Do you think you are still looking out the window?

N: We're excited. I know we saw it and we're starting to follow it, but we can't believe it! The light seems whiter and bigger now.

B: Does it still seem to be overhead?

N: It's overhead and in front of us, but it feels like we're not in the same place where we were. I feel like he is scared all of a sudden.

B: Mike is scared?

N: Yes.

B: Is he saying anything?

N: He is saying something, but I don't know what.

B: You are aware that he seems to be scared, but what are you feeling?

N: It feels like we're out of the car now, standing beside it and looking at this thing.

B: Is it still there?

N: Yes and it's really slow and hovering.

B: Has the car completely stopped now?

N: Yes, but I think the headlights are on though. Mike is on his side of the car and I'm on my side, and we're looking up. It stopped moving and now we're on a different road that's a lot smaller. There is tall grass or stalks of corn along the side of the road, but nobody is around.

B: Is the object still there?

N: Yes, but it's not moving.

B: Has it come any closer than it was before?

N: It's very close, maybe 100 feet. It feels like there is a focus on him. That's really sad, but I don't know why.

B: It makes you sad to think that?

N: Yes. He is wearing a black shirt and it's about him this time.

B: Now that he is out of the car and looking up, is he scared?

N: I'm just watching and I don't feel like I'm involved, but maybe I am. I think the light is on him, that white light.

B: Is there a light shining directly on him?

N: Yes and his body seems to be twisted.

B: Take a look at that light. Is it just an area of light or does it have more of a shape to it?

N: It's a beam and it has shiny particles in it, but I don't know what it does.

B: Is this beam coming right down to him?

N: It's on him where he is on the other side of the car. I'm not moving very much over here in the dark. I'm not standing in the light.

B: Do you think you can move? Have you tried to move?

N: I don't think I can. I'm leaning on the car. I feel bad for him because he is so terrified.

B: How do you feel over on your side of the car?

N: I feel kind of dull and numb. The only thing I can feel is feeling sorry for him. I want to help him, but I can't.

B: What makes you think you can't help him?

N: I don't know. I'm in the dark on the other side of the car and I'm not a part of it.

B: Does it seem to be focused only on him?

N: Yes, but you know what? I think there is somebody behind me in the tall grass because I hear a rustling. I don't know what that is. I feel there is some activity going on around both of us suddenly. It looks like he is frozen in this position. He looks very weird.

B: Do you see anybody with him or do you just see him in the light?

N: Some of them are coming out in a circle around us all of a sudden. They're curious about him, but I think they know me.

B: They are all focused on him?

N: Yes.

B: You don't feel that any of their attention is on you?

N: No, but I feel like they know me and I'm just waiting. I don't know what they're doing with him, but I'm waiting.

B: Just notice what does happen now, whatever it is that you can see or be aware of in any way.

N: I think he is just gone for a while and I'm outside waiting. Somebody is waiting with me.

B: Okay, now let's just slow it down here and go back to that point where they're all around and it seems like their focus is on him. Just notice anything that happens that causes you to realize that he's gone.

N: Somebody touched me on the shoulder and made me stop moving. I don't know what to do. I wasn't present for a minute and now he's not there anymore, but the ship is still here.

B: Is the beam of light there any longer?

N: No. It's misty and grey out, but still a little lighter around me somehow.

B: Is there is a light around you now as you are standing next to the car?

N: Yes.

B: Is the whole beam of light gone now?

N: Yes.

B: Now that he is gone, do you notice anything that's happening for you?

N: I don't like being on a road by myself like that.

B: Are you completely alone or does it seem like somebody else is there?

N: Somebody is here with me, one of them. He is shorter and smaller, behind me, and keeping me calm and patient.

B: How does it seem like this one is doing that?

N: I think he touched the inside of my arm or my wrist. I think it has something to do with my body endorphins relaxing or sedating me. I can feel those long fingers.

B: What do they feel like?

N: Rubber.

B: Now this is summer, so is your arm bare? Can you get the sense of that feeling?

N: Yes, my arms are bare.

B: How would you describe what you feel?

N: It's a little creepy. They move so slow and deliberate. It has something to do with the inside of my right elbow, pressing it with his finger.

B: Can you tell just by the feeling of it how many fingers there are?

N: There are three fingers and a thumb is behind the other side of my elbow.

B: What did you say the texture feels like?

N: I might be kidding myself.

B: You might be kidding yourself about what?

N: About being where I am. I feel the metal.

B: Well, let us just notice anything else.

N: I might be inside.

B: We're talking now about the kinesthetic feeling, but you can also notice whatever you see, visually. You have been feeling the three fingers and the thumb on your elbow. Just keep on feeling and noticing visually.

N: I might be inside and seeing him ahead of me. I don't know if this is real.

B: Just put aside any evaluating and notice what you do see.

N: All right then. I'm inside a dark round room and I see Mike on a table.

B: What does the table look like?

N: It's a metal table. I see him without a shirt on and he is really fighting, really freaked out. Oh God! I think they're sticking something in his penis!

B: Well, just go with what you are seeing. You see him on the table. Is he sitting on the table without a shirt on?

N: He is lying down, wanting to move and he is freaking out.

B: Do you see anybody else with him or is he just by himself?

N: I see a few of them at the foot where his legs are. It's dark at the edges and there are some lights over him on the metal table that look like stainless steel.

B: Now we're going to break this down into little increments, just moment-by-moment. You are seeing a few of them at the foot of the table. Describe what they look like.

N: I have never seen them move like this before. Their legs and arms seem particularly skinny and dangly. It's like they're moving a little bit in slow motion and their heads are enormous with little, teeny holes for the nose. I can barely see the mouth. I don't know why, but they seem taller than usual.

B: They seem taller than what?

N: They're like the others I've seen, but just a little taller for some reason.

B: What do their eyes look like?

N: Black.

B: How about the shape?

N: Big, almond, and black. They're very intent on him and they seem bothered that he is so upset. They're running around. Oh, man!

B: Notice what they seem intent on and if it makes sense to you.

N: They're doing something to his genital area, putting something long into his penis. Yuck!

B: Can you get any sense of what that is that they're putting in there?

N: It looks like a glass tube. It's a thin, long glass tube with a round bulb thing on the end.

B: Does it seem like the glass tube is empty?

N: There is some liquid in it, a clear liquid in a very thin glass tube going somewhere.

B: How is he reacting to this?

N: Oh, he's terrified.

B: So it seems like he's conscious and aware?

N: He's calming down now, but at first he was flailing all over with his legs, trying to hold them together.

B: They're trying to hold his legs apart?

N: Yes.

B: Does it seem like they're still doing this procedure right now?

N: They've taken something from him and put it in a little clear, round case.

B: Is that something a fluid?

N: It could be. It's white and cloudy and it could be ejaculate floating in water in the round tray. It's a clear tray and the top comes off it. I think he's tagged with something, a computer chip in his foot. They're doing that, too. Wow!

B: At this particular time, what does it seem like they are doing with his foot?

N: They did something to his foot, his arch, or his heel. They put something in it and then they made the skin close up by putting a light on it.

B: Just notice now. You can slow this down if you want to. Replay it to get more details.

N: I can see the hair on his legs and his right foot.

B: Where do you seem to be? Are you are getting a good view?

N: He is on the table and I'm next to him, sitting on some kind of chair. Someone is with me and holding my elbow. That's it.

B: Now that feeling of someone holding you on the elbow, does it feel like it did down on the ground?

N: Yes, it's the same thing and the same being.

B: Your right elbow?

N: Yes, they're pressing inside. They press the wrist, too.

B: At the inside of the wrist?

N: He took his fingers from inside my elbow and pressed three fingers from one side of my wrists and the thumb is underneath. There is something specific about the pressure.

B: Do you have a sense that they know exactly where to press?

N: Oh, yes, they know everything.

B: And the effect of their pressing? What effect does that seem to have?

N: You don't move. You're paralyzed and you can barely feel your body. It numbs you. I don't think it is so much emotional as it is physiological, stopping your body's response.

B: Does it stop your emotions, too?

N: I'm very calm, but that might be from something else. I'm mentally here with him and disturbed that he is going through this, but I don't feel my feelings.

B: Do you know if he has ever had experiences like this before?

N: No. I don't know anything.

B: Do you think that Mike has any sense that you are right there?

N: No, I don't think he is paying any attention to me.

B: This is a very wonderful opportunity right now. Would you like to find out what they are doing to him and why they are allowing you to see this? You may have other questions, as well.

N: Well, I wonder if what they did to me gave me the endometriosis I suffered with for years. I would tell them to leave him alone! Why are you doing that? Oh! He turned his head and looked at me.

B: So you are asking this of them?

N: Yes, and he has turned his head. He seems curious.

B: So, one of the beings turned his head?

N: Yes. He's at Mike's right leg, and he turned around and

looked at me. There is somebody in the background, one of the big white guys standing in the shadows. He's the leader. I know he was there from the beginning. You know, when you imagine yourself filled with white light, feeling surrounded by love, that scares them off sometimes. It really strikes them oddly, puts them off, and makes them stand back.

B: Maybe they're not so in touch with the white light.

N: No, they are, but they just take the energy and use it mechanically for manipulation, not for evolution. I use it for the evolution of my psyche and my soul. When you have the emotional side as a human being you understand that the power of the light affects the soul, but they can't use it that way because they don't feel that part of themselves. They're dead to it so they can't apply it in that way. No wonder they want what we have, but they can't get it that way!

B: Okay, so we're now withdrawing slowly from that whole scene with him, you and with the beings. We are going to bring our attention back to the car. Just notice how you returned to the car that night.

N: It's very dark. The lights are off and the car isn't running. We're sitting in the car and it's as though we fell asleep.

B: Oh, you're already sitting in the car?

N: We're sitting parked in the dark. Mike's head is thrown backward and his Adam's apple is sticking out.

B: What is your first awareness? Does it seem like you are completely awake or are you slowly coming to.

N: I'm very awake, but he's not awake so fast and just coming to. We're not talking and just driving back to my house. We told everyone we had just seen a ship. My parents thought we were nuts.

(End of Excerpt)

JUNE 1991 INCIDENT

Summary of Event

The following event was Nadine's first *consciously recollected* ET encounter that took place in Oak Creek Canyon in Sedona, Arizona in June 1991. After the event, Nadine was fully conscious and immediately able to recall a good deal of the experience. She described the abduction, in person, to Barbara Lamb on January 23, 2005, and provided her recollection, in writing, as follows.

It was June 1991, and my friend, Pamela, and I took a trip in a van to Sedona, Arizona. We planned to spend our time between camping and staying at a motel, but on our first day when we were about twenty miles outside of Sedona, rather than continue to our planned campground, we pulled off onto a winding, dirt road in a very desolate area. Although there were no signs posted, we traveled a bit down the road until we came to a rustic campground that had about ten small clearings that served as camping spots. Several of the cleared areas had picnic tables, but there was no electricity or other facilities available, other than an outhouse. A small creek ran along the side of the whole property. It was dusk and no other campers were in the area and we were alone.

Pamela and I unloaded the car and set up our camping equipment: an ice chest, lantern, chairs, food, and other items. We each had bags containing our personal belongings that we placed on a picnic table not far from the van. Then we built a small fire and sat talking under the stars until about 11:00 P.M. Although there was no evidence, I had an overwhelming, uncomfortable feeling that we were being watched. Our sleeping bags were in the back of the van and I left Pamela to go to bed a little after 11:00 P.M. Shortly thereafter, Pamela came into the van and went to sleep, too.

The next thing I knew, I was sitting up, wide-awake and outside of my sleeping bag, shaking badly. My heart was racing and my skin was sweaty. I was in a terrified state of mind, feeling that I had just been dropped back into the van from somewhere else. My mind kept racing with terrifying thoughts and images of having been taken from the van against my will by someone not human. I clearly remembered that the back door of the van had suddenly opened, swinging upward, and a small, thin arm reached in to pull me out. I was shocked to see that the hand only had three fingers, instead of four, and the skin was grey! I was dazed and unable to move on my own until, suddenly, I was being moved along the ground somehow by two little creatures, one on each side of me. I could remember that something had repeatedly hit my face as we moved along, but I could not turn my head to avoid it. Ahead of me, I could see an extremely bright, white light and we appeared to be getting closer to it. Then I could remember nothing more until I felt myself being dropped back into the van abruptly. I looked at my watch and it was 3:30 A.M. I could not shake the insane feeling that someone was coming back to get me again. Nothing vaguely similar to the event of that night had ever happened to me before and I was completely panic-stricken. Although I made every attempt to wake Pamela up, even shaking her, she remained in a deep state of sleep.

I stayed awake for the next two hours with my face pressed to the windows just watching, fearful for their return. About 5:30 A.M., just as it was beginning to get light, I was finally able to wake Pamela up and we exited the van. I shared my experience with her and she listened attentively as we walked about the area. It did not take long to realize that, although nothing in the camp had been touched, including her possessions, all of my personal belongings that I had left outside were gone. We searched the wooded area thoroughly, fifty feet or more in every direction, but were unable to find my bag, towel, clothing, and shoes. Knowing how ridiculous it sounded, I nevertheless tried to reassure myself that some animals must have taken my things.

I was so deeply disturbed by the incident and unable to shake my fear, that we packed up immediately and left the area. Although we mentioned the incident briefly several times on our long drive back into California, I wanted to simply forget what had happened and put it all behind me.

Over the next six months, I became obsessed with the idea of moving to New Mexico, a state where I knew no one. In January 1992, I finally gave in to my strange obsession and moved to Santa Fe. Once there, I immediately began to experience nightmares of strange beings and I developed a host of what I considered irrational fears. Much of the time, I felt that I was being watched, particularly when I felt compelled to go driving on some of the more remote stretches of roads around Santa Fe. Many times, I would wake up in the middle of the night, certain that someone small was in the room watching me or outside my bedroom window. Often, at any time of the day or night, I would become so overwhelming sleepy that I would suddenly fall asleep wherever I was, and reawaken two or three hours later. I would often find myself in a different location or my body would be in an odd position. There seemed to be nothing that I could do to avoid these episodes. Over the next few months, other strange behaviors developed. I found myself constantly checking for someone or something lurking in closets, behind doors, or in the back seat of my car. I had frequent, strange nightmares of seeing beings, examination tables and ships, and a fear of the dark developed.

Although I refused to trust my memory of the Sedona incident, I was so distraught by my irrational fear that I finally went to see a psychotherapist on two occasions to discuss it. I only shared with her that I believed something had traumatized me on that summer evening in the campground. Using hypnosis, she easily succeeded in putting me into a deep state of mind for about fifteen minutes. As she regressed me back to the incident, I became so agitated that I was crying hysterically and flailing my arms and legs wildly. During those few minutes of recall, I did regain additional memories that helped me to fill in some of the blanks from that event.

I saw that when I was taken into the clearing toward the white light, it was a metal disk-shaped object. I could only see directly in front of me, so it was difficult to be certain, but the craft could easily have been 80 to 100 feet in diameter. Once inside the craft, I was forced to lay naked upon a metal examining table. At this point in the regression, my emotions escalated quickly and the therapist moved me past the scene until I was calmer.

I recalled one other aspect of the encounter when I was standing in a large room and I noticed an opening into another area. I leaned over to look through it and saw what looked like other human beings lying down in rows of flat containers as if they were sleeping. Someone pulled me back from the doorway and my next recollection was a physical sensation of being dropped. Then I heard a big swooshing sound of air and I found myself back in the van, wide-awake and sitting upright.

Abduction in Sedona

Excerpt from Hypnotic Regression May 5, 2005

B: Ok, we are now back in the year of **1991**, on or about June 15th. You and your friend are on your way to Sedona, Arizona. After driving for many hours, it's getting late and you decided to pull off the road into a campground for the night and you went to sleep in the van. Tell me what happened then.

N: She and I were sleeping. They opened up the back door and took me out.

B: And probably the van was locked, I would imagine?

N: Yes. I felt it coming before it even happened that night. I had a dream of an alien one month before we went. Then I needed to go to Sedona.

B: So, you are sleeping in the van with Pamela, mid-June 1991, in Sedona, Arizona.

N: I'm sleeping on a pad. It's warm inside, but cold outside now.

B: What position are you sleeping in?

N: I'm on my back.

B: At this moment that you are reliving, is Pamela in the van with you?

N: Yes.

B: And does it seem like she is sleeping, too?

N: Yes.

B: Okay. So, you're sleeping on your back in the van. What are you aware of now?

N: There are shuffling sounds outside.

B: Does it sound like the noise is coming from down toward the ground or is it higher?

N: It's right where I am.

B: Is it coming from just one side of the van or more sides?

N: I don't know.

B: And as you are hearing the shuffling sounds, are you thinking about what that might be?

N: I just want to sleep.

B: Even if you are falling asleep, a part of you can be very aware of the shuffling sounds. Just notice them. Your mind is aware and has recorded everything that has happened even when your body is asleep. We are allowing that to come forth now, the awareness of the shuffling. Notice if it seems to be in just one very small area or if it's more extensive than that.

N: It has scattered around me outside, and around the end of the van.

B: Just let this experience continue now. You are hearing the shuffling and notice if something else seems to happen.

N: I hear a click on the back door and the door swings up, way high. The light is on now, too.

B: The van light inside?

N: Yes, the lights go on when the back door swings up.

B: So the light is on. Is it dark outside?

N: Yes, it's very dark.

B: So, you are lying there on your back.

N: I'm not on my back. I sat up when I heard the click and the door opened. I sat up.

B: And your eyes opened?

N: Oh, yes! This is freaky. This has never happened before!

B: So you see the light on and the darkness outside. Is the back door of the van all the way open now?

N: Yes. The three fingers are awful. It has three long, skinny fingers, a skinny arm, and it's reaching in the van towards me. They have little pads or cups at the end of the finger. It's a weird color, kind of a grey-green.

B: The skins of these fingers are grey?

N: Yes.

B: Do you see what those three fingers are attached to?

N: No, not right away I don't. I just see the hand and I don't ever remember seeing anything like it before. It wants to pull me outside.

B: Has it touched you at this point?

N: It must have touched me because I'm out of the van, but I don't remember getting out. The door is wide open and I'm standing on the ground in my nightgown. Oh God, I'm scared!

B: So you are standing on the ground now and you're terrified. Can you notice if anybody else is there with you?

N: I'm barefoot and I see dirt and little pebbles underneath my feet. I feel frozen and I can't move at all. I'm paralyzed!

B: And yet you are upright?

N: Yes, I'm upright and standing.

B: Is anything propping you up?

N: No. I'm just standing up straight and no one is touching me. There is a little taller one in front of me and three shorter ones on each side, but they're not touching me. They're just holding their hands underneath mine, cupping them, but not touching me. They make me rise off the ground a few inches. I can't move! It's like being tied up. You can't move at all!

B: So you are not very high off the ground?

N: No. just an inch or two. I'm surprised I'm not cold.

B: Oh, I'm surprised too. Now, the one you mentioned who is in front, can you get a look at this one?

N: I don't remember him like this before, but he's taller, definitely bigger. The others are little, something like three and a half feet tall. This one is five feet tall at least, maybe even taller. He's not like the others.

B: When you say he is not like him, does he have a different kind of head and face?

N: It's very angular. He is more double-jointed or something and brown. It's as if all the others serve this one.

B: What about his eyes distinctly? What are they like?

N: His head is different. Wow! It's as if they work together. I thought he was a Grey like the others, but he's not like that at all.

B: Is he with the little Grey ones?

N: Yes, he is.

B: Okay. Well, that's interesting to know.

N: He is guiding the way, telling them where to go and how to do it, and they just follow him. They're just like little automatons. I thought, at first, that there was just them and us, but there's a lot of more of the little ones around us just watching. I can see a face in the bushes watching.

B: And are they like the little ones who were around you?

N: Yes, They're very small, with big heads, forehead, and big eyes. This guy in front is gross! I don't know what he is. I think he made himself look like the others at first.

B: What about his body?

N: It seems like his skin and his body is very rough and crinkly. There is some kind of armor over his chest. It looks like it's external,

Small Grey Being

as if it's a plated armor attached to his body. I don't even get this at all! The way he walks is weird like his legs and joints are facing backward.

B: What else is he wearing besides that plate of armor? Does he have any other clothes on that you can see?

N: Where the joints are, it's very knobby. He might be wearing something like the armor on his legs, too.

B: So when you think of the joints, you think of elbows?

N: His knee seems like it's backward. Gross!

B: What about his sense of weight? Is he heavy, medium, or very light?

N: He's bulgy in spots where the joints meet, but more slender on the sides. He's ugly, not like the others. The others are not ugly and there is nothing really to see. This one is ugly!

B: And what do you sense about this one?

N: He is in charge, guiding and directing the thing. It doesn't seem like they think for themselves. I'm glad I'm calm. I can't feel the terror now.

B: Has anything like this ever happened before?

N: No.

B: Have you seen these little ones before?

N: At home when I was little.

B: Did you remember them in the waking part of your life?

N: No.

B: Okay, but is there something familiar about them that you are experiencing right now in Sedona?

N: Yes.

B: This taller one who is so different looking, is he familiar in any way? Do you think you have ever seen him or one like him before this experience in Sedona?

N: I don't know. He's a warrior, a fighter. The others are smooth and they're not very conscious. This guy is much more powerful, but he is not like the guy on the ship. The guy on the ship was tall and pasty-white color.

B: What about his shoulders? Can you see his shoulders?

N: He is bigger on the top and smaller on the bottom

B: Okay. Does he have broad shoulders?

N: Yes. He could be as much as six feet. His skin is so bumpy, with ridges and very rough looking.

Tall Reptilian Being

B: What do his hands and fingers look like?

N: He is much more aggressive and he feels everything, and there's some kind of odor.

B: Just notice that now.

N: Yes, there's some kind of odor, and I think he is playing himself down so he doesn't look so threatening or so big. He seems to be curious about me. He's directing us toward a big white glow in the woods. You can see it in between the trees. They don't touch me, just put their hands under my hand and it lifts me.

B: They don't pick you up or anything?

N: Nobody is touching me, but they're still moving me.

B: So, all they do to move you is gently touch your hands?

N: No, they're not touching my hands, just putting theirs underneath. They cup their hand underneath mine and it raises me a couple of inches.

B: Are the little ones doing this?

N: Yes. They're moving me away from the van.

B: And as you are moving away from the van, is the back door of the van still up and the light on and everything?

N: I can't see behind me. I'm looking forward.

B: Okay. Are you seeing the woods ahead of you or are the woods all around you anyway?

N: We're in the clearing where we were camped and now I'm just scooting across the ground. The tree branches are hitting my face, and I can see this white light ahead of me. It's so bright on my eyes!

B: Are you still upright?

N: Yes, standing straight up. I don't know where the other guy has gone. Now we're in another clearing and the white light is so bright.

B: And what do you see now with that brightness?

N: I feel lots of presence around me, although I'm not seeing them right now. I'm feeling so detached from my feelings, but I think this is the scariest thing that ever happened to me in my life!

B: Yes, because you have no idea what's going to happen.

N: The light looks very misty now, not so bright anymore, just a fading mist. I see metal and it looks like brushed stainless steel. You can't see anything on the metal, and then I see a crack all of a sudden. It wasn't there and then this crack opens up and it slides apart so you can go in.

B: Have you been able to see the shape of this thing?

N: I only saw a piece of it, a soldered edge where it comes down. It's definitely like a saucer, round. I can see it going around. It has to be much bigger than what I can see. It has to be a hundred feet across. It's huge!

B: Is it shiny at all?

N: It's not shiny, more like a dull, brushed stainless steel that's silver color.

B: And then suddenly it sort of opens up?

N: It does, and I'm moving in. I don't want to go in.

B: Are they moving you in?

N: Yes. I see a guy on the other side of the room and I do not like him. He turns and looks at me with a look that says I'm nothing. He is tall, maybe seven feet tall and skinny, with some kind of long coat or cape on. Oh, I mean nothing to him!

B: Is he anything like that taller one who was with you outside or is he different?

N: No. He's different. This one's very pale, pasty-white, much straighter, and smoother.

B: You mean a straight upright posture?

N: He has weird eyes, too. There is an instrument panel in front of him. The ceiling comes down to the bottom of the room and there's an instrument panel there, too. There are lots of different things in that center section.

B: Can you notice anything about the instrument area there?

N: In front of him is an examining table that reminds me of a slab at a morgue. I know I'm going to end up on that table!

B: You know you are going to end up on it?

N: Yes.

B: Does any of this seem familiar at all, like you've been in this kind of a room with these unusual beings and seeing that table and being on that table ever? Does it ring a bell?

N: When I was very small, maybe five years old.

B: Are you remembering from way back, years ago?

N: I know they hold your wrist and then somebody is assigned to you just to keep you calm. Their whole concentration is just to keep you calm. They do something with pressure points on your wrist, too.

B: So it's a very specific way of holding your wrist?

N: Oh, yes!

B: Are you feeling upset now?

N: This is just not good because this is all very sexual. It's a real bad situation to do that to me, to hold me down and touch my body. All of its ugly, bad sexual things. Terrible!

B: At this moment as you are thinking that and reacting, are you on the table at this point, or are you standing there just remembering? What is happening actually?

N: I'm on the table and there's a hose going up into my vagina. There's three of them down there and they're very intent, pulling something out of me.

B: Do you feel any discomfort or pain? Are you feeling any of that?

N: I feel some cramping in my abdomen.

B: Describe the hose.

N: It feels like it's up to my belly button. I think it's a vacuum.

B: Is there a sense of pulling something out?

N: Yes, there's suction to it.

B: Now, we can slow down this experience, no matter how fast they're going with what they're doing, so let's slow it down and look at the three guys there. First, do you have the impression that they look similar to each other?

N: Yes, they all look the same. They look like the little grey ones, but a little bit taller. Maybe it's because I'm lying down.

B: That's right. You have a different perspective.

N: They're taking something from me.

B: What comes to mind? What do you think they're taking from you?

N: It makes me think of abortion.

B: So, they're taking something from you now. You had mentioned earlier that they assigned somebody to you to keep you calm. Is there anybody there with you right now on this table with this procedure happening?

N: The same one is still on my right arm. She's there. I don't know why I think it's a female, but maybe because she is in my head.

B: What does she seem to be doing?

N: She is just doing her job, keeping me calm. There's no complication to it. She just has her one simple focus and that's for me to stay calm.

B: Is she communicating anything else to you?

N: She's making me feel it's okay and that everything is all right. She wants me to believe that nothing bad is going to happen to me. That's it.

B: Is she saying these words aloud?

N: No, they don't do that.

B: Are you saying that she is speaking telepathically to you?

N: Yes.

B: Does this calm you down and help?

N: Oh, it works. You'd be out of your mind, screaming and fighting like crazy if they didn't do that to you, but I'm paralyzed and I can't move. They try to make you believe it's the right thing and it's okay. There's nothing okay about it.

B: Yes, but they say things like that.

N: They try to make you believe it's natural, but it's not. Nothing is right about it. Nobody asked me!

B: So you have all this influence coming to you suggesting that it's okay, but you can see through it right at the same time?

N: Yes. I don't believe it. I know when I'm with perpetrators. It's all so unbelievable!

B: Take a good look at this one to your right. What does she look like?

N: It's funny how they blur everything, but when you look hard, you get something different.

B: Yes, that's true.

N: I think she may not even be a Grey. She looks like a mixture of something and has that female look to her. Wow! I wonder if I know her! It's interesting, but I get the feeling she doesn't want to be there either.

B: Do you think she is a human female or some other kind of female?

N: I think they have done something to her. I don't know what she is. She's a mix of something. It's not a good thing.

B: What kind of thoughts come up for you about that?

N: She is a mixture of something, yes.

B: You said you also thought does not want to be there either?

N: Yes. It seems like slavery, in a controlled society where everything is controlled. It's not good or kind.

B: Do you have the sense that all of these beings, whatever they are, are doing this routinely without any agenda?

N: There is an agenda. Of course, they want something.

B: And right now, concerning you, what do you think their agenda is?

N: Right now, it's just physical information. It's always about information.

B: So, whatever they're doing now to your body, taking something out of your body with a suction instrument, is for their purpose of learning?

N: You know what else? They're also fascinated with my breasts!

B: Are they examining you there now, too?

N: They put something into my right breast.

B: This is a very personal thing for you. How are you reacting to them doing this to you?

N: Well, I'm dramatically subdued. My terror was overwhelming, but now I feel so subdued, like being shot full of drugs. They want to watch progress all the time, to see how things are progressing.

B: Specifically, is it your breast or your reproductive organs that they are particularly focusing on?

N: Oh, both.

B: Do you have any sense that they're talking or communicating with each other in any way as they examine you?

N: I can't tell because my emotions are so bad, but there may be communication going on with someone to the side. It's so shocking, so unbelievable! I don't ever remember anything like this. It's all you can do to accept it.

B: Which, of course, is what they want you to do.

N: Yes.

B: So, with all of the events happening now, are you saying that it doesn't seem as if these kinds of procedures have ever happened before to you? Or, do you think that this may be sort of familiar to you?

N: I may remember something from when I was little.

B: There is some vague kind of sense or vague memory?

N: Believe it or not, but I think my father is here, too.

B: When you say he is here, too, is that strange for him to be here in this environment where this stuff is happening?

N: He's somewhere nearby. I knew it happened to him, too. I knew it did, but I didn't want to believe it.

B: No, and probably he doesn't want to either. Do you have the sense that he's there also as a victim?

A.   Yes. He's completely out of it.

B: So, you are both having a very unusual and invasive experience.

N: I never thought of him being here at the same time. Gee, that's strange.

B: So, with his being there nearby, do you have a sense that it might even be in the same room?

N: Yes, he is off to the side.   They're curious about the relationship. I have to get out of here now!

B: Yes, we are going to leave it now because of the time.

(End of Excerpt)

FEBRUARY 23, 2000 INCIDENT

Journal Entry February 24, 2000

I had another weird experience that seemed like a dream of abduction, but I *know* it wasn't a dream. I must remember every detail; I told myself I would when I was there.

I was standing outside at night with a man. The sky was very dark and the stars incredibly bright and alive. I saw that the moon was out and then I saw what looked like a red light that came between the moon and us. It was very large and you could see details on it. The man was about to go away and I was going to go back inside when I realized that something was happening overhead. I called to him to come back and we watched the sky, mesmerized. We saw them flying through the sky and I knew they were coming again. Their formation was something incredible.   There was a mother ship with several smaller ones following behind in formation.

Ships in the Sky

I remember my knees buckling beneath me and I found myself sitting on the ground in shock. They were coming again for me. I went into a house and there were people gathered there who seemed very stunned. I thought of going to hide but instead decided that I would stay as sharp as possible and remember as many details as I could. I wouldn't go into the trance this time. I would stay clear and ask questions. Suddenly they appeared in the room and there were three kinds - a very small Grey, a taller white and what looked like a third kind, transforming from a white? There are lots of coming and going and I'm on the ship now. They don't want me to write or to remember. I'm standing around watching the beings and lots of movement of humans, too. They're working with others and not paying too much attention to me.

A machine is near me, five inches wide and it opens up to measure and calculate something in my body. I'm at a table sitting and I see glass tube cylinders on the table, devices for measuring, analyzing, and calculating. I see symbols sheets with curves and angles that they want me to memorize or test me with, or put ideas into my brain. I don't want to and I'm mentally fighting them. I feel like I'm able to stay more aware than I have ever been before. I tell the one across the table that they're not God. They may think they are, but they're not. He tells me that my concept of God is not accurate, that I should expand my mind, and that God is not permanently one-way. He says that we're the focal points and that God goes outward into infinity and expands as we expand, that I should not limit my thinking. I ask the one across from me at the table if he is surprised how awake I can stay now. He tells me (thinking), yes, and I ask him why he thinks that happened. He says they did not count on me falling in love, that somehow that affected my psyche and woke me up more. He tells me also that they're here because my father died and they wanted to see how that affected me.

There is a hole in the back of the chair. I feel someone behind me implanting something, probing my back through the chair. Is he helping my body or harming it as an experiment? Did he do something to my head, too, and the left side of my neck and ear? I know one of them is back there where I'm sitting. He is trying to work on my low back without me knowing he is doing it. It hurts and I'm trying to pull away, but something is keeping me tied to the chair.

The one across from me seems agitated by my resistance but he (seems like a male) is preoccupied with working on some instrument. The one is to my side seems female, and is leaning over me, setting things down on the table. Someone came over and set down this small, round tray of insects in front of me. There are other trays on the table too, filled with these insects, but mine look different. I see one that looks like a centipede and I think they want me to eat them, but it's disgusting and I don't want to think about that. The female to my right looks somewhat human and I realize they're trying to alter themselves to look human now. Is she really a Tall White or a Grey? I asked her in my mind if we repulse them and she responded telepathically, "Yes," and nodded.

Then the next thing I know, it seems like some time has passed, and the items and the bugs are gone, as well as the being who was on my right side. The symbol sheets are still in front of me and they still want me to look at them, but I keep asking questions in my mind. Then they show me what looks like a ship or huge capsule of silver, matte metal that's spinning and coming down from the sky at great speed. It hits the desert floor and keeps spinning, digging into the earth and burrowing below the ground. Somehow, they show me what it looks like as it goes deep into the earth and opens up. It keeps spinning but ends up looking like an octopus with lots of huge arms extended around the center. They show me a city off in the distance from where this lands, and then show me a scene of a modern home and kitchen appliances.

Testing at Table

Excerpt from Hypnotic Regression January 22, 2005

B: We're going to do a regression back to the night of **February 23, 2000**, when you were closing up your father's house. He had just died recently. We are going back to February 23, 2000, at about 3:30 A.M. when you suddenly came to and found yourself sitting upright on a mattress on the bedroom floor where you had been sleeping. In this experience, you are awake and aware that you are sitting up, experiencing heart palpitations and sweating. You were so emotionally affected by the experience that you wrote in your journal for over an hour. This is the experience that we are going to go back to now. Tell me, what is the first thing that you're sensing as you're sitting on the mattress in the bedroom?

N: I feel like someone's outside the window, at the side of the house.

B: Are you on the first floor of this house?

N: Yes.

B: You have the sense that you actually see this, or are you just sensing that someone is outside the window?

N: I sense that someone is outside the window.

B: Do you have any particular feelings about that?

N: It's like what happened in Santa Fe. I would feel them outside the bedroom window where I was sleeping.

B: As much as possible, be there completely now. Be on the floor lying on the mattress when you are having this awareness. Notice the position that you are in as you are having this awareness.

N: I'm not lying.

B: What position are you in?

N: I'm sitting straight.

B: Are you leaning against anything? Are your eyes open at this point?

N: My legs are folded underneath me and I'm sitting cross-legged.

B: Does that seem perfectly normal to you, that you would be sitting up like that?

N: Not really. My back feels straight and rigid.

B: Does that take some effort? What else are you aware of now besides your position?

N: I feel as if I'm being watched.

B: From which direction?

N: From behind.

B: And how are you reacting to that?

N: Irritated. I feel damp, cold, and irritated.

B: And the dampness and the cold feeling, is that something that you had before you went to bed on that mattress?

N: No. I'm feeling metal now, pressed against my back like a straight-backed chair with ribs across it.

B: So, is it pressing into your back?

N: I'm pressed and flattened against it, almost as if I'm tied to it.

B: Emotionally, how are you doing with this?

N: I'm very irritated and very angry. It's like the first time, but I'm just fed up and tired of it.

B: Is this reminding you of something that has happened earlier? Is there something familiar about this?

N: I'm just tired of it happening and tired of being pushed around.

B: Are you saying that you think this has happened before? Is that right?

N: I know them. I know the guy across the table.

B: Have you changed your location at all? You're talking about a guy across the table? What are you aware of now? You have been sitting up in the mattress and you have felt this thing press against you.

N: I'm not on the mattress. I'm sitting at a table that feels like the tabletop is glass, but underneath is stainless steel.

B: And how about the temperature of that?

N: It's warmer now.

B: So, you're in a different location it seems at this moment and not on the mattress in the bedroom anymore, right?

N: Yes, and the tabletop is glass with a three-inch lip around the side. It's thick and smooth glass.

B: Is it comfortable to sit on?

N: I'm not sitting on the table. That's in front of me. I'm sitting on some kind of chair. My back is pressed against something and it's cold. In front, its smooth glass and I can see the table legs underneath. They're wider at the top and get smaller at the bottom.

B: How far in front of you are you focusing?

N: It's right in front of me. I can't move and my legs aren't folded. They're bent at the knee.

B: Are your legs touching a floor of some sort?

N: I can't really feel my legs very well, but they could be bending normally and I'm sitting.

B: Are you feeling any discomfort in any way?

N: My low back aches. I want to touch the table because it looks so soft and so smooth.

B: Are your hands touching this table?

N: Yes, and when the glass comes to an end, it's like round at the end. It's perfect and very smooth.

B: Is there any color to that?

N: No, everything is very pale and grey, clear glass, or metal. This table is almost pretty.

B: What is the lighting like?

N: The lighting seems to come from everywhere, but I can't tell where it comes from. Everything is lit up, but not brightly. Then there's a separate light in the center of the table, underneath the vials and tubes. There's a light shining up from under the table, too. The cylinders have liquid in them and each one is at a different level. Wow! It makes me think of a musical instrument.

B: Are these lined up next to each other?

N: They're in the center of the table, a bunch of them placed in a circle. Some are tall and others short, but they're all filled with different amounts of clear liquid. I don't know why, but it seems like it has something to do with hearing or sound. The guy across from me is the same one who was across the room from me the first time in Sedona, the one with the funny head that looked like an insect. The pasty-white guy is in the background that I don't like and he's annoyed by me.

B: The one in the background looks familiar to you?

N: He looks very familiar to me and we don't like each other very much.

B: Let's just pause for a minute and take a good look at him. What does he look like?

N: I wonder why it's so hard to look at him. I feel like I can't move my head up to look at him, even though he's right in front of me, looking at me so intently. He knows that I'm determined to remember and I'm really going to be present. That's all there is to it!

B: Good.

N: There's a female here to the right of me and someone else behind. The guy across from me wants me to do something, and I don't know why it seems to me it has something to do with my hearing. I still can't look at the one across the room, but I know I've looked at him before when he was a long way across a room.

B: Okay, but he is closer now, right?

N: He's much closer.

B: Now, I wonder if it could be anything about his eyes that's somewhat off-putting that you don't want to look at. What do you think it is that you don't want to look at?

N: He's not like the other ones and it seems like his eyes are more on the side of his head, with little sections to the eyes.

B: Facets, sort of?

N: Yes. Facets and little bumps in the eye, and his head comes out more

B: His face?

N: Yes, a different shape that's broader at the top and then goes in and out again at the chin. He has a mouth and eyes like a fly, not black. They're reddish or brownish-black with separations, like little cells in the eye.

B: His skin?

N: It's pasty-white.

B: Concerning where you are sitting, does this one in the background seem to be about your height or does he seem to be shorter or taller?

N: He's bigger and much taller than I am. I think he's the same one that watched me from the first time. He has that same look.

B: Do you think he is the same type of being or do you think he is the same individual? Can you determine that?

N: I think he's the same being, the same one from Sedona when I walked in the first time. He was standing on the other side of the room and seemed so emotionally cold. He's cold now, but he's more curious and irritated because I'm fighting now.

B: What is it about him that makes you feel sure that he is irritated? Is he showing it with facial expressions?

N: No, I can just feel it because he's trying to do something in my head and I'm resisting him, fighting him a lot now. I'm not afraid like I was, so I'm annoying to them. I kept thinking that when I got older this was not going to be happening anymore.

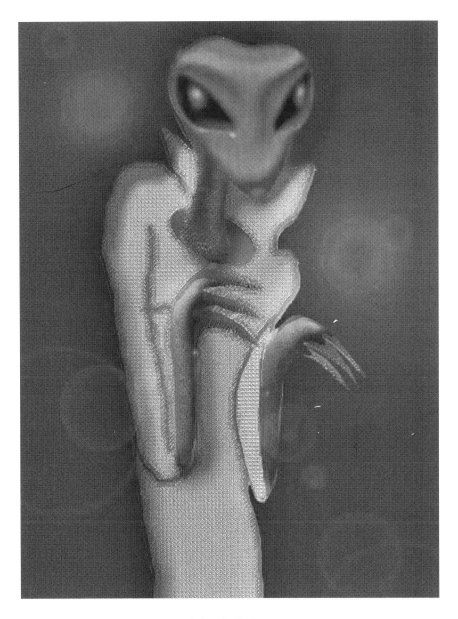

Mantis Being

Tall White Being

N: He's not like the other ones and it seems like his eyes are more on the side of his head, with little sections to the eyes.

B: Facets, sort of?

N: Yes. Facets and little bumps in the eye, and his head comes out more

B: What is it about your age that you think would mean this would not be happening now?

N: Just because I'm in menopause.

B: Are you surprised because being in menopause does not seem to make a difference to them? Does your reproductive functioning have anything to do with how you feel about this?

N: I think it happens in stages. For ten years, they treat you one way, and in the next ten years, you're studied another way. Every ten years, there's something different.

B: Are you just talking about your regular cycles as a woman?

N: Yes, that's how they break it up.

B: Do you have a sense that they have studied you in these ten-year cycles, and that that has been going on for quite a while, longer than you thought?

N: I think it's a part of my life and it has been a part of my father's life. They're more curious about me now, though, because I'm resisting and I'm irritated. I'm not so scared anymore and I hate it all. It's not fair! Someone else is here now, to my left, a female that I have seen before. She seems like she is in her early twenties and they want to see how I act when I see her. She is drawn to me and standing by me now.

B: What kind of feeling do you get emanating from her?

N: I feel sorry for her. She has seems lost somehow and not free.

B: How about describing her?

N: She is tall with thin, wispy blond hair on her head like baby hair, cut short. Something is not right with her skin, or it's so transparent you can see through it to the veins. Her eyes are very pretty and different with big eyelids and lashes. I also remember having had a profound dream of seeing her when she was a child.

B: She does have lashes?

N: Yes, and she has lips. She's pretty.

Adult Hybrid Female

Hybrid Female as a Child

B: Does she seem to be registering anything now you are looking at her? Is she looking at you?

N: She was looking at me, apparently fascinated by me more than I realized. I think she wants to touch my shoulder.

B: Would you say that she's a human being or is she somewhat different from that?

N: Something is not right with her. She's detached and sad.

B: Does she seem to be registering anything now you are looking at her? Is she looking at you?

N: She was looking at me, apparently fascinated by me more than I realized. I think she wants to touch my shoulder.

B: Does she touch you?

N: She is standing next to me touching my shoulder and the edge of the chair. I know she has just been out on a few occasions, but she never stays very long and I don't know where she goes.

B: The one you were just looking at with the blonde hair, did she seem tall?

N: Tall like me, but not extremely tall.

B: Tall like a regular human woman?

N: Five foot seven or eight.

B: Have you been aware all through this?

N: Well, they do something here to keep you emotionally calm when they do this.

B: So they press into your wrist?

N: Yes, press it, or make energy go into the area. I remember the first time that they did that, the being was little and she never left my wrist. She was there the whole time. I want to touch this thing on the table, to look at these symbols in the clear case. It closes and then when you open it up there is a bunch of them in there, not just one. All the transparent sheets are hooked together at the bottom like the spine of a book. Each symbol has a meaning. At first, it looks as though you could poke your finger through the shapes, but you can't because the design is covered with a thin, iridescent film of some kind. I'm supposed to flip through the sheets to learn from them, and I'm tempted to touch them.

Symbol Sheets

B: They want you to?

N: Yes, absolutely. They want me to remember these, but I don't understand the meaning of the symbols. They're supposed to tell me something later. If I remember them now, then I will remember them later when I see them. They're all pointy and each one contains information. Altogether, it has to be memorized.

B: And they're watching you as you're handling this?

N: Yes, they're watching in front of me. The thing that I thought looked like a centipede in the clear, round tray is metal. It's in the glass tray with a lubricant. The top lid is a magnifier so it makes the thing inside look bigger. I just realized that I'm not alone at the table. There are others like me.

B: You mean more humans?

N: Yes, and I think the table is longer than I thought and I have just been focusing on one end of it. There are other humans here, but they don't have the symbol box. They're doing something with everyone. I thought that they wanted me to eat that thing, but now I don't think so.

B: What thing?

N: The thing in the tray that I thought looked like a centipede. I can see that it's a lot smaller with many little legs on it like a little comb. They don't want me to eat it because it has something to do with my ear. They kept it in this liquid to keep it clean. They distract you by having you look at something while they're doing something else to you.

B: Are you in front of it? What's on the table?

N: The box of symbol sheets is on the table.

B: You do have two things to concentrate on, right? Both the case with the sheets and symbols and the other tray of liquid with the object in it?

N: They're doing something with the little metal object. All I have to do with is look at the symbols sheets. I know they're trying to put the information into my head that they can get out later. I think the film in the center of the symbols has something to do with the information. There's more meaning to the symbols than just their shape. Each one also contains information within it like a computer chip. The film is pearly and reflects pale colors like a rainbow.

B: Would you say that it's pretty?

N: Yes, like sheets of geometrical shapes that you use for drawing. I'm wondering how these glass tubes have anything to do with my ears or measuring my hearing. I think I'm being watched and there are three other people here at the table, too.

B: Does it seem like you are the only one looking at the symbol sheets?

N: Oh, yes. I'm looking at the symbol sheets and they look at me when I look at the symbol sheets.

B: By *they*, are you meaning the other types who are not like you?

N: Yes, they're watching, me but there are other humans at the table, too, who aren't a part of this and they seem like they're asleep. Maybe they test one at a time.

B: How many beings would you say are watching you as you are looking at the symbol sheets?

N: There's the one in front and across the table, the girl to the right who's supposed to keep me calm, and someone is behind me doing something to my back, but he's not paying attention to that thing on the table. He doesn't care and he's just manually doing something. There are three or four of them also standing behind me and they're exchanging thoughts as they look at me and watch in my head. They see in my head and then they communicate, but I can't tell what they're thinking and I can't hear them.

B: You can't tell what they're thinking?

N: No.

B: How is it that you know that they're communicating with each other?

N: I can feel it in my head. He wants them to notice me. Yes, I think he likes me, after all.

B: Does it seem like he's in favor of the way that you're responding to these sheets of information?

N: I think that when you do the right thing it surprises them and they don't like it, but they respect you.

B: Okay. Do you have the feeling that you're being more aware of what you're doing than the other ones at the table?

N: Oh, yes.

B: Are they in a trance or whatever?

N: Oh, yes.

B: And they know that you are more aware, right?

N: We've talked about it. They think some kinds of emotion make us stay awake more and they like to watch that happening because they think it affects our perception. They want to know what emotion does to perception because they're not like that.

B: Do you think that your emotion does something to your perceptions?

N: It can bring you closer to something and give you extra insight into what you see. It gives you a foundation to hold on to if you are scared, but if you're too scared, it shuts everything out. If you're passionate, in a positive way, it wakes you up and you see more.

B: So, some emotion is helpful.

N: It is energy. I guess its strength.

B: Do you think that it's your emotional energy that makes you remember these experiences?

N: I want to know what they're doing. I don't trust them and I never have. I don't like the way they do it and I want to know what's going on. Finding out and knowing gives you power and strength.

B: You mean the things they have you doing here?

N: No. It's not just this. This is nothing. This is easy.

B: Compared to?

N: Some of the stuff is bad and it's just not right. If I have to be bothered with this stuff, I'm going to watch and see everything and learn. I don't know, but I'm not afraid anymore.

B: Good.

N: Sometimes when you're angry, it's good.

B: It's strengthening, isn't it?

N: Yes, it can be so long as it's not irrational.

B: So right now in this situation, by the table and these other beings and everything, you're not feeling scared?

N: No.

B: Good. Apparently, you're really paying attention. Is that right?

N: I do now.

B: Do you feel mentally clear in this setting, sitting by the table and the information sheets and noticing what you notice?

N: Yes, but I'm still fuzzy. It's like a cloud pulled over you and you have to strain and push to see. You have to work hard for clarity.

B: Sure. You're doing some of that, aren't you?

N: I'm seeing more all the time.

B: So at this table now, you're straining to see what's happening?

N: Because I do work so hard to see, they consider me more interesting and valuable. Some push through and some don't.

B: Do you feel that these sheets of information are making sense?

N: No. I think that if you look at it you're remembering it even though you don't realize it. I think you're getting the information stored in the brain and, later on, when you see something, you'll know it. I'm not fighting it. I'm cooperating because I'll use it, too.

B: I suppose there's a chance that maybe they want you to do something in the future. What sense do you have about that?

N: I was standing in a dark room, and there was this picture on the wall.

B: In their territory, would you say?

N: Oh, yes. It was very dark in that room and they were showing bad things on the wall and they wanted me to look at them. Things were blowing up, you know? Something is going to hit the earth.

B: Is that the sort of thing they were showing you, that something's going to be hitting the earth?

N: It's going to get here. Maybe something is going to hit.

B: But it's difficult to look at it?

N: Yes, like looking at war, but it's not war. It's just things falling apart and collapsing, blowing up with lots of smoke when something hits us that came from somewhere else.

B: Is it a frightening thing to look at?

N: Not really. I'm so tired of being afraid of everything in life. I know everything is going to be okay. I trust in the order of things and I trust in God that it's all part of a process and I'm only here for a while anyway. There's that question, though, about how should we prepare? I have food and shelter, sleeping bags, and water ready and I'm waiting. I bought my SUV car because of the big disaster so I could pack my supplies. I'm getting ready.

B: Do you think that maybe some of that could be stimulated by the scene that you are watching on the wall, as well as what we hear right here?

N: I don't know. I think about survival a lot, about being prepared so I can take care of myself. I want to take medical classes and CPR to be ready. I want to be self-contained, mobile, and ready.

B: Let us go back to sitting at the table with the guy in front looking at you and those behind, talking about you. See if you can tune in more to what the ones behind you seem to be saying to the one in front of you and vice-versa.

N: I think the one across from me is trying to convince them of something about me. It's as if those beings behind me are so mentally connected right now that they're almost operating as one entity.

B: Does that seem to be the same type? Can you tell me?

N: I can't tell because they're behind me.

B: And you can't turn around?

N: I tried to turn my head, but you can't move in this seat. I don't know who they are, but he's showing me to them and they're very skeptical. It's almost as though he's proud of me, although there's no emotion as if I'm an experiment.

B: Would you like to know more about what the one in front of you is trying to convince them about?

N: Yes. Why am I here? What are they doing? What's with the symbols? What do they want me to do? Why would they want me to have this?

B: Right now, they're aware. Certainly, the one in front of you is aware that you're not just a regular person. He's aware that you have awareness, so this is your opportunity to ask them your questions with your words out loud or just with your mind. Focus on the one in front, and ask him any question you want.

N: Okay. I want to know about something when I first met you. How old was I when I first met you?

B: The important thing, I think, is to project that to him directly and then just be receptive and see what you get.

N: Two. I get two.

B: Okay.

B: And ask for more if you want more details about that. You were age two.

N: I want to know if my father was involved. I feel like he is saying, yes. I want to know if something has been put in my body for recording. I want to know how they find me.

B: Okay. Let us stick with one question at a time. Is there something in my body?

N: Okay. Is there something in my body?

B: And if so, what?

N: What's in my body and why is it in my body? You know, the visual that I have right now is that he has shifted his vision. Where he was looking above me before, now he's looking right at me. He tells me five.

B: It's as if he's saying five?

N: There's five. One has these little things on the bottom, little dots, and fibers that touch you, but it's smaller. There's also one that looks like a centipede and another that's black. There's a long, thin implant in my nose! What else is there? There's a round, silver implant. This makes me think of all the body stuff and it's a turnoff!

B: You mean things that have been done to your body?

N: Yes, and it's disgusting. I don't want to think about it, but they can monitor everything from your temperature to everything else going on in your body. It's amazing!

B: Do you think they're doing that right now?

N: I think there's a whole record of everything. They can see changes when your emotions get intense and it gets their attention. It alerts them that something's happening that's triggering intense emotion, maybe because intense emotion affects blood pressure. Who knows what else it does in the brain! There may be something at the base of my skull and the base of my tailbone.

B: What is it about that at the base of the neck?

N: I think that when they examine you and put that instrument into your colon, it serves as an x-ray device and has nothing to do with a person's sexuality. It's like our MRI but from the inside.

B: They certainly seem to have great interest.

N: I wonder about the chakras and if they have some way of recording disruptions of the energy meridians in the body. Maybe that's why they don't get sick because they know how to change their energy. I have so many questions.

B: Focus now on another question that you might want to ask them.

N: What's going to happen to the earth and when is it going to happen? I'm seeing that something cataclysmic is going to happen that comes here from the outer atmosphere. Is it an asteroid? I think a huge asteroid or another race of beings that are on their way here that will knock the earth off its axis.

B: When you ask that question now, just wait a little bit more and be very receptive, because you have ways of picking up. You're tuning in more and more.

N: What's going to happen on the earth? It feels like tsunamis are part of it when something comes here from somewhere else and hits us. I feel like I'm supposed to be prepared and help others to be prepared.

B: Does it seem as if these beings want you to be prepared?

N: They want me to be prepared in a way that serves them. They don't necessarily want me to be prepared in a way that serves me. They're not concerned about me. It's not about me. It's about the planet. They don't care about me except that I'm a curiosity and something that will serve them. The others are still asleep and slumped over.

B: The other humans over there?

N: Yes.

B: But, you're aware and you're not slumped over?

N: I don't know what happened here, but it seems like somebody was trying to prove something.

B: The guy in front of you.

N: Yes. They're surprised how awake I am, very surprised.

B: We're going to have to conclude this, but just ask anything else now that you want to know that you would be particularly interested in.

N: Okay. I will ask the one across from me. What do you think of me? He's intrigued by my fighting spirit and thinks I'm more assertive and direct than most of the humans they work with.

B: Could you imagine him smiling or showing any facial gestures?

N: No. They're not like that. They're not colorful and they don't seem to have a range of personalities. It's just missing. If you don't have emotions, you miss a whole other dimension, so they're missing a lot and they can't seem to manufacture it.

B: So, they seem to be studying you and your reactions and the chemicals that flow through your body and internal body reactions. Do you think they are collecting a lot of data?

N: Oh, yes. I think they identify with the physical aspect and how the organism works, but they can't identify with how the dimension of emotion affects an organism. They don't understand how perceptions are colored by the emotions and how they affect the physiology. They can't grasp how and why things stir us emotionally, so much that it makes our heart rate jump. In their reality, storage is based on what they've experienced, not what they're imagining. We're existing in an almost multi-dimensional state and they can only grasp one of our dimensions. Of course, they're in another dimensional state where we can't grasp all of their dimensions. It's like there's a piece of us that's similar in the manifested form where we can meet, but each of us exists partly in another realm where we can't meet.

B: Do you think you explain that to them?

N: Yes. I try so hard to conceptualize the emotional condition and make it more tangible, something you can touch and measure more. I think they like that.

B: So, maybe you haven't been aware of this before, that not only are they sort of studying you and taking information from you, but that you're giving them something.

N: Once, I had this experience when I was with them when I was really, angry. I was speaking, not just thinking then and telling them they weren't God. They can't find us the way they're trying to and they couldn't comprehend it anyway because of its evolution and they haven't earned it. They responded in a way that seemed surprised to me, and it's the only time I ever saw that. I kept trying to think of the word, *hologram*, and as soon as I thought of the word, a hologram appeared over the table in an instant.

B: From your thought?

N: No, I don't think I did it. I was trying to explain emotions to them and a hologram might best describe the multidimensional aspect of emotions. They knew what I was thinking and showed me one. About a week later after that experience, I bought a book called, *The Universal Hologram.*

B: So, there was a real response there.

N: Oh, yes! It was a huge hologram and they liked the analogy.

B: Wow! That was an interesting experience, wasn't it?

N: I don't know why, but it made me feel good; I made a point.

B: So, sometimes, and in that instance, there's real two-way communication, isn't there?

N: Oh, there was on this occasion.

(End of Excerpt)

FEBRUARY 1, 2003 INCIDENT

Journal Entry February 1, 2003

Damn! I woke up with full details again of a so-called dream. They have given me information regarding how they will be arriving, and they are coming in a very big way. When they come out into the open on earth they will first go to the desert areas outside of big cities where they will install huge metal energy systems underground. These systems will arrive in the form of a spinning capsule coming from the outer atmosphere. When it hits the ground, it will cont64inue spinning, digging far down into the earth's crust, eventually opening up to extend many huge arms, looking somewhat like an octopus. It will settle into one spot and continue to rotate very fast underground emitting some kind of energy force. They suggest to me that this gigantic equipment will be able to affect the brain functioning of hundreds of thousands of people at the same time in a nearby city. These huge mechanical devices might also be installed on some of their ships, acting as a transmitter, disrupting the mental processes of humans in the area. They might also use this wave of energy that puts our brain into that fuzzy, hypnotic state in other kinds of equipment, as well, but they're not susceptible to it themselves.

Now I realize that I was a part of a group of humans in the room and a leader was there, too. He was having all the humans come over for an examination of some kind. They seemed to be checking our height as we stretched up onto our toes and then stretched outward with our arms to have them measured. He knew that I know there is something they want me to do for them in preparation for their great coming. He was in my mind, telling me that the little wars on our planet are nothing and that we have no idea what big really is. He also said the microwave came from them. I feel like this experience was taking place on a military base.

Massive Underground Hypnotic Device

I was either in a room and looking out of a window or standing on some part of the base's landing strip. I could see their smaller crafts move away from the mother ship and descend onto the landing strip. Human aircraft were on the ground and the alien inhabitants were boarding the human aircraft to take over their control. I watched as most people went into a trance-like state, but I was still able to maintain some degree of awareness. I think the anger that I feel from the knowledge of what they're doing sets up a resistance in my mind. I'm committed to remembering all that I can from every experience and I will keep spiritually strong to fight them.

My environment changed again and I felt as if I had been transported into a house in the desert, next to where one of the giant capsule machines had burrowed underground. I knew one of the beings in the room and he seemed to have a mild curiosity about me. The Whites do know how to make themselves look like humans, and they can manifest out of thin air, or they seem to. They may be traveling through space by using dimensional revolution, which allows them to stop time, using physical ships for part of their journey and then dimension travel for a part. It appears as if they're just materializing, but they just make it appear that way to us. I think they may be here before we see them. (I swear that I can hardly believe it when these memories and thoughts come out of my mind!) I'm not afraid of this weird stuff anymore, only angry, really pissed off! These are things that I clearly remember have seen:

- An air force strip with air force planes being stopped and boarded by non-humans.

- An area where a huge energy transmitter is installed which affects whole neighborhoods, allowing them to numb us, invade our homes, monitor us, and send signals to ships in the outer reaches of earth's atmosphere.

- Rooms that look like a kitchen where they appear, but the rooms are not kitchens.

- Landings in a desert and seeing a whole family affected by the big underground machine.

(Additional note—I told Pamela about the sound I heard several nights ago, just before I fell asleep. I heard the clicking noise as an actual audio sound with my ears. Could I have resisted then? Was it a sign of their coming?)

Someone told me they are watching us and when they come, it will be very dramatic. By 2027, they will be here full time, moving about out in the open, but they will not have yet gone into many of the small rural communities. That's where we will build the resistance.

(At the time of the printing of this book, no regression had been conducted regarding this event.)

DECEMBER 20, 2003 INCIDENT

Journal Entry December 20, 2003

I dreamed of UFO's again last night, over Chatsworth. There were a dozen or more very huge ships that became transparent at times, as though they could appear and disappear at will. I noted one that was projecting a massive light beam into the sky. They were perhaps a half-mile up in the sky and many were dropping something large and square to the earth. These huge squares were attached to huge, semi-transparent parachutes that looked somewhat like a jellyfish. Many people were terrified as they watched the objects fall from the sky, and some were leaving the area to go and hide. I tried to pack up some personal items at my home but had great difficulty keeping my wits to gather my gear. A woman appeared whom I didn't like or trust. I sensed that this being was a male who was creating an illusion by projecting himself as a woman. This was a confusing and uncomfortable dream.

(At the time of the printing of this book, no regression had been conducted regarding this event.)

MAY 1, 2004 INCIDENT

Journal Entry May 3, 2004

Ok, so I will note it down, the fact that the night before last I again experienced the alien encounter phenomenon. I had another experience that masked as a dream. They were coming again and I tried to hide in a sleeping area that seemed like a bunk. I wonder if it was on a ship. Of course, they found me. An alien officer, dressed in a red suit (disguise), like I have never seen before, told me I was one of several humans who they have worked with for five years, and that a very big event will happen in the not too distant future. We will be called upon to act as liaisons between the aliens and humans. They told me that my psychic powers would accelerate and, as they sat me next to another human, I could hear him talk without talking (telepathy). The man was about 60, very upset and fretting about what was happening to him.

I remember they showed me a large number of ships of different sizes, twelve or more, up in the sky at night. It led me to think that a variety of alien species would be coming to the earth. Blah! How I hate this phenomenon, whatever it is! That night there was blood on my sheets and pillowcase. I had a bloody nose in my sleep, again!

(At the time of the printing of this book, no regression had been conducted regarding this event.)

AUGUST 1, 2004 INCIDENT

Journal Entry August 2, 2004

I woke up abruptly exactly at 6:01 A.M., August 1, 2004. It was a blue moon (full moon). Last night I was abducted. It felt like I was driving my car and suddenly realized that I had missed my exit. I thought about taking the next exit but then realized that I was actually out of my car and watching someone else going off the freeway without me. I now recall being removed from the car and seeing another woman with marks on her body that were also on me, large scrapes or marks across the front below my shoulders from something they had done to me.

Next, I was in a large, crowded room filled with a mixture of beings, different kinds of aliens, and humans. I was getting more and more lucid as time went on. Something occurred that I think was more examinations. Then, I was sitting at what seemed like a large rectangular table with several beings, perhaps six or seven different species. They were talking with me about things, with a particular interest in the *way* that I thought. They made me think about my brother and indicated they were interested in the close relationship we have, in that we came from the same two parents. Somewhere during the beginning of this contact I recall calling out to God to help me, and their response seemed almost like amusement, in their limited fashion for humor.

In my mind, I was trying desperately to get them to understand that although we humans may have inferior intellect and brain development compared to them, we do have another dimension that they seem to be without and that's the emotional dimension. I tried to find the right words to explain this, as I believe these beings want to believe that emotions are something to be dismissed as only interesting, and a frivolous, unimportant curiosity only. I tried to make them understand it's a much bigger and more important aspect of evolution, that it's more of a hologram, a deeply dimensional and much more complex development of the human race than ever appears at the flat surface. As I was struggling for the word *hologram*, to describe human emotions, they helped me to mentally find the word by instantly displaying a hologram over the table where we sat. It appeared over the table as a large three-foot square shape that looked like glass and moved sideways where you could see it was a block. This image hovered over the table for perhaps 10 seconds. They seemed very impressed with my analogy. I also tried to explain to them their lack of compassion and their inability to relate to another living being other than analytically.

I remember sitting at this table across from a large white being that seemed like the leader or supervisor, who asked me if they had ever impregnated me. In my mind, I told him that I had been pregnant. Then he seemed to note something down, but I couldn't see. He thought, almost aloud, "Then we did." In the next instant, they showed me two babies sitting on a flat cushion through a wall behind him in another room. One of the babies was staring right at me, blankly, yet seemed to recognize me. I kept thinking that they were doing this to get my emotional reaction. I think they were making comparisons between my brother's development, my babies, and me, having all come from my parents. During this experience, I also recall having a sexual desire for one of the male aliens present. I think that some of them are male and female and that they've done many genetic altercations to create many species.

All of this took place in the abduction dream state. My memory is extremely vivid, and it doesn't seem recorded like a typical dream. In fact, in this particular abduction, I was so incredibly awake and lucid that I remember this now as though I've just walked out of that room a minute ago and I'm recalling the event. Besides the fact that they wanted to see the effect of the blue moon on the human body, they were also curious about my desire to laugh tonight. They wanted to see the physical effect on my body from having laughed so intensely and for so long. (Earlier in the evening, I had gone with a friend to a movie with the sole intention being to laugh as much as possible because I needed an emotional boost. I think they were somehow aware of that fact.)

I believe some extraterrestrial races may possess a limited emotional capacity and, therefore, might covet what they consider to be a greater emotional awareness in humans. There have been instances when I observed an extraterrestrial displaying what seemed to be disdain toward one or more humans, while at the same time showing curiosity and interest beyond a clinician's perspective. I have also recognized situations when an extraterrestrial demonstrated a favorable attitude toward humanity, to the extent that I felt they would like to help us. During this episode, several ETs appeared quite amazed by an analogy I presented, and they were surprised by how awake I was. I kept thinking while it was happening, "I'm here! I'm really here and I'm awake. This is really happening!" Also, during an interaction with a female Tall Grey, I remember asking her in my mind if they found us ugly and she responded affirmatively.

During this experience, I also remember a female with smaller eyes and short, light-colored, fuzzy hair. She looked at me curiously and I thought that she was almost pretty, with her larger lips and eyes that had a blue iris and white around it. Seeing this female stirred me deeply. She appeared to be in her early twenties. Strangely, I kept thinking that she was my daughter, taken from me when I was thirty or thirty-one! She dressed very differently, too. I wondered if she could help us or if she would even want to. I just looked at the clock and its 6:54 A.M. I have been writing nonstop for more than fifty minutes. To remember this amount of information with such clarity is astounding to me.

(At the time of the printing of this book, no regression had been conducted regarding this event.)

DECEMBER 30, 2004 INCIDENT

Journal Entry December 30, 2004

Okay, I got it – 4:44 is a return time! I woke up abruptly with another fake dream. Another abduction took place! It began with a screen memory of my being in a small house with a wooden floor and a famous actor was in the room, too, in a very foggy state of consciousness. The room was dark and it sounded as if music was playing, a haunting and slow melody. I went up to the actor (they know I like him) and we danced. He was very disoriented, and so fascinated with him that I didn't realize at first that I was being taken. The music stopped and I don't remember anything until I become aware that I was with this man in a large warehouse. He was scared, terrified actually, and I was struggling to stay alert, but not succeeding very well. I knew they were coming and we tried to find a place to hide, but then I lost consciousness.

I woke up and found myself lying on a cold surface that seemed like a shelf. The actor has disappeared. One of the beings was on the right of where I was lying and I could see a new fine line scar on my right inner forearm, three-quarters of an inch long, that was never there before. The next thing I knew, I was in a room with the actor again and I knew they were coming back. He kept mumbling a name that sounded like, "Lisa! Lisa!" I thought it might be his wife. I kept telling him to touch something metal, that the physical contact would help him concentrate to maintain more awareness. I tried hard to comfort him. I sensed he was from the San Francisco area. (What were they testing me for?)

Then I found myself in what seemed to be a huge train station or hall with many people waiting. They must have one of the beings specifically assigned to you (in this instance, two?) to maintain your drugged, foggy state by their mind manipulation. I remember standing and waiting for my turn with two of them fading in and out to my left. They were looking like aliens one moment and then fading into humans. I realized then that they're among us, using some kind of mass hypnosis to make us see them as human. Still waiting in line, I heard a man screaming and saw one of the taller ones doing something to the man's mouth and I was next. For some reason, I wasn't afraid.

I kept thinking about trying to get back to find the actor's room, and then I suddenly saw my brother on the other side of the room. I knew it wasn't him, but one of them pretending to be him. He was very aware of me as he slowly got closer. I didn't like this one; he was exceptionally emotionally dead. Next, I remember being in a room and telling someone that you can maintain your awareness with practice, mostly by fixating. I saw a material thing, a table, waving in its form as they tried to make it look like something else. (I learned the next day that the actor who appeared in this experience is married to a woman called, *Dina*, not very dissimilar from *Lisa*.)

(At the time of the printing of this book, no regression had been conducted regarding this event.)

MAY 20, 2005 INCIDENT

Journal Entry May 21, 2005

I woke up this morning very emotionally uncomfortable with a vague, weird memory from last night. It may have been a dream, but I don't think so. It could have been something else, but it was definitely an alien abduction. It was with a species that's very different from any of the other extraterrestrials I have seen before. It's a foggy memory so I can't remember very much except weird sounds. But still, for some reason, it's really bothering me.

Excerpt from Hypnotic Regression May 28, 2005

B: We are going to go back to May 20, 2005, when you were in bed at night and we are going back to the unusual dream you had that night and to the unusual sounds you heard in the room on the same night when you had that dream. You have some memories of the dream and we're going back now to see exactly what occurred on that night, Friday night, May 20, 2005. Tell me, Nadine, what's your first sense that something is happening?

N: There's someone at the foot of the bed, the right of the bed.

B: As you are aware of this, does it seems like your eyes are closed?

N: My eyes are closed, but it feels like someone is standing there. Could it be possible that still another type has come along? These are not like before. They just step out of nowhere and appear. It's as if the atmosphere just opens up and here is this being, looking at me curiously. He has a big head and a skinny body, and he's tall, white and pasty. He looks like a bug, but he's not as tall as the extremely tall ones. This one is about five-foot-five and translucent like you can sort of see through him. His skin is very thin and has a bluish tint to it. Boy, he is really looking at me curiously! I'm wondering why this species just all of a sudden develops some interest when you've never been involved with them before.

B: So, you're aware of them and you're wondering about this.

N: I just realized that there are three of them and I think they're physically in the room. They're pulling the covers back on the bed! Oh, this is scary! Something is very different this time and it's not at all familiar to me.

B: Is it just because they're different that this seems especially scary?

N: Their agenda is different from the others. They want me to do something or they want to program me to do something for them. They're pulling me out of bed.

B: How are they pulling you out of bed?

N: I think that they physically moved the covers back and I just floated up out of bed. Whoa! This is just much more dramatic. Is this happening or could this be my imagination doing this? This has never happened before!

B: Just keep on letting it happen.

N: Well, I'm floating right up out of the bed and through the walls. Gee, I've never had any memory of anything like this ever happening before.

B: Well, let us just keep on going with this.

N: It feels like - whoa!

B: So, you're floating up out of the bed and through the wall. As you are going through the wall, are you going through it alone or does it seem like they're with you?

N: I don't think they're with me.

B: So, it seems like you're going through the wall on your own?

N: It feels like I'm going through the wall at an angle, up and out through the apartment to the side. I'm moving towards a very bright light. Oh, God!

B: Well, just keep going. You're moving through the wall into a very bright light.

N: The bright light is pulling me upward.

B: Does it seem like you're outside?

N: Oh, yes. I'm outside. This is a physical experience!

B: Okay then. Your whole body is there and doing this.

N: Yes, and I don't know them. Everybody does it differently, but this is very different. I'm cold.

B: Do you have a sense that you are still rising?

N: Oh, yes. I'm rising into a shape. My logical mind doesn't want to accept this! I must be the one creating this! Their blue skin may not be skin. It might be something covering their skin. I think they're more like humans, not the way we look, but their brain may be more similar to ours than the other beings, in the way it functions.

B: Okay. Are you seeing them again now?

N: Yes, but I've been pulled up into this thing in the air on this beam of light.

B: What position are you in?

N: I'm standing almost straight up, but it's as if I'm flying.

B: So, there's no effort on your part and it's just happening to you?

N: Yes. You just get into the beam of light and that's it. It just pulls you up, but I can't tell if I'm actually physical. Still, my whole body was in the bed. If they weren't taking my actual body, why would they pull the sheets back on the bed?

B: Are you aware of being able to move at all?

N: No, I can barely move and my body is twisted, but I don't know why.

B: Okay. Just keep on going with this and see what happens.

N: All right. I don't think they're very interested in humans or me in general. I think they're interested in another species that's visiting us and they want to know what's been happening with them. Isn't that interesting? This seems more like an interrogation of sorts and just an intellectual exchange. They want to know what I've been doing with the others. It's like there might be some kind of competition between them here on earth. I can see now that these beings have weird arms that are double-jointed.

B: So, they seem to want to know what the other species have been doing with you.

N: Yes, that's exactly it! This species is more primitive, and I can tell they're more emotional because they seem agitated. They're not as detached intellectually either as the rest of them, and it seems like they may have a more barbaric or combative nature. They want information from me about the others.

B: So, it seems that they came to you for some information on that?

N: Yes. I don't think these beings have the same degree of intellectual development as others do.

B: So, they have emotion?

N: They have a wider emotional range, at least. They know the sensation of really wanting something. I think they have emotional reflexes built into them and I think they're calculating.

B: Are they very urgently trying to get this information from you?

N: Yes, they're making me see that they want to win and not lose.

B: Do you think it's a competition going on between them and the others?

N: Possibly, but they want to monitor my interaction with them.

B: How are they going to do that? What's happening with you right now? What kind of position are you in?

N: I'm sitting in some kind of chair. It looks like a dental chair with a hook around my neck to hold me. There are lots of lights around me, too. I don't think that these beings can read your mind as clearly as the other ones, and they're communicating with sounds. They might do some telepathy, but I think they mainly use some kind of clicking sound to communicate. They vary the pitch from high to low and they speed it up or slow it down.

B: Does it seem like they're talking to each other with the sounds?

N: Yes, and they move funny with jerky movements. I can see their heads better now and they look like ants.

B: What do their eyes look like? Take a look at them.

N: They're big and their eyes are on the sides of their head. They look like the head of an ant. I don't think I've ever seen anything like them before.

B: Describe the color of their skin.

N: The skin color is white on the surface, but there's some brown and blue that shows through, too. Overall, it's a weird thing.

B: Would you guess that they're about five feet or so?

N: Yes, that's close.

B: How many of them are around you right now during this questioning or whatever they're doing?

N: There are five of them, with three in front of me. You know, they definitely can't get into your neuropathways and mess with you like the others. I can tell that's frustrating to them.

B: Can they just look into your eyes?

N: No, they don't have that same kind of power. There's a sense of urgency about them and they seem emotional.

B: So, you have three of these beings in front of you, but where are the other two?

N: One is behind me to the right and the other one is to my left. I feel like they're monitoring me with devices, and that they depend on the mechanics of the devices for their information.

B: Do you see these devices? Are they holding anything or using any instrumentation of any type?

N: The one to the left has a cloth cover or drape over him. There's some kind of instruments, but I can't seem to focus on them at all. There's part of an instrument that's extending from this chair and focused on my head.

B: Although they're making the clicking sounds, does it seem as if they're talking to you in any other way? Are they asking anything of you?

N: They don't seem interested in connecting with me. It's more as if I'm carrying something that they want that records information

B: So they're not asking you questions, but maybe they're using instrumentation somehow?

N: They're mainly interested in having me do something for them. They don't want to know about me. I'm just carrying something for them.

B: So, remain aware now. Can you get a sense of what they want you to do?

N: They want me to get information about the other visitors and they're picking up those of us who have been involved with them. This is like real star wars!

B: Just keep noticing anything more about what they want from you.

N: They just want to use me to record information when I'm inside to see what they doing with us.

B: Do you have any feeling as this is going on that somehow they're right now getting information from you in some way?

N: I don't know. Maybe I'm wrong about this.

B: Do you have any sense of feeling their thoughts? You got a sense that they want information.

N: They just made me see a ship entering into the water. I think they're trying to ask me questions by showing me pictures, or they're making the pictures in my head.

B: What does the ship look like that enters the water?

N: It's a round ship, but I don't think these beings use round ones. Their ships are older and more cigar-shaped. These ships are coming down from the atmosphere and entering the water at an angle. That makes a lot of sense. There could be ships under the ocean.

B: Is this is a new thought for you about ships coming down into the water at an angle?

N: Yes, but now I keep thinking they're off the coast of California, northward, around Santa Barbara. They want to know if I'm aware of that, but I didn't know that.

B: Are they still showing you that picture or did they show it to you only once?

N: I'm just going to tell you what I'm remembering. It doesn't make any sense to me, but I'm not going to analyze this.

B: Just let it roll.

N: All right then. I being told that there is a military base up there on the coast somewhere, maybe navy.

B: Up towards Santa Barbara - that way?

N: Yes, right on a strip of the ocean. I can see one of the saucer-like ships going into the ocean alongside where that base is.

B: Do you get a sense if anyone at the base is aware of this?

N: I don't know, but they want to know if that's where I have seen the people in military uniform in the last year? I think they know.

B: You think they know what?

N: I think the government knows and these beings know that.

B: You think the government knows about these ships going into the ocean?

N: Yes.

B: Just notice if they're showing you anything else.

N: Forests. They're showing me rain forests. They may be able to put pictures in my mind more than I thought. I wonder if they put things in teeth.

B: So, you wonder if they put things in your teeth.

N: Yes.

B: What is it that's coming up for you?

N: I think they leave something in your tooth that picks up different frequencies and information. They hide things in the metal in your teeth and you don't know it's there.

B: What are you aware of now?

N: It's just like espionage and covert activities.

B: Are these guys showing you the covert activity of others?

N: Yes.

B: Does that mean others, like other civilizations?

N: It means other civilizations that are not from here. I just really don't want any of this to be going on. I don't want any part of it. I'm tired of it.

B: Does it seem like what they're showing you involves you in some way?

N: They're trying to instill in me that the planet is going to be threatened and at risk in the future by another race that doesn't care about our planet. They're trying to tell me that some races are good and another one is evil. They want me to believe that they're genuinely concerned about the earth and that the others are only concerned about experimenting. Somehow, these beings seem more present and genuine, maybe not as bad as the other ones that create all the dream sequences. It doesn't seem like they're creating as much illusion. They just don't feel as complex as the others. I also notice that these beings have mouths.

B: They have mouths?

N: They do have a mouth and I think they eat with it. The others do not eat at all.

B: Are you seeing that now in action?

N: You can just tell.

B: Have they done anything to harm you in any way?

N: I don't think so.

B: Have they been doing any physical procedures with you?

N: Certainly not sexual, but they may have done something with my head and my neck or the upper part of my body. I don't think they have any interest in reproduction issues or any of those things at all, although they might hide something on me so that they can get information. That could be happening. I can't see if it's going on because it's masked somehow, but I don't feel threatened. I just feel angry and annoyed. I was scared at first because they were so different and their approach and the way they travel are different. I still struggle when I hear this kind of stuff come out of my mouth.

B: Well, let's just move forward now to the very last things that you're aware of in this experience.

N: They're communicating and trying to come to some kind of conclusion and there may even be some dissension between them about how to approach things.

B: And are they making those clicking noises as this is going on?

N: Yes, off and on. They're expressing a great deal of emotion. The other beings show curiosity, but they don't express any emotion.

B: Is there anything you would like to ask them while you're with them?

N: Well, no. I don't want to linger. All of this annoys me and I have to process it all. Wow, I'm coming back now and it's fast. I hear a whooshing sound. It feels like I've just come back through the air in a different way and fast.

B: And when you say back, back where?

N: I'm back home.

B: Back into your bed and everything?

N: Yes.

B: And back in your bed in your room now, what are you aware of? Are you actually in bed, under the covers, or how is that?

N: I'm sitting in bed in the dark and looking at the clock and its twenty minutes after three.

B: Are the numbers lighter than the clock, so you can see them in the dark?

N: Yes, it lights up in the dark. Boy, this wasn't a very long experience at all.

B: Do you have a sense that you elected to come back then or do you feel like they choreographed that and they brought you back?

N: Well, I feel like I was physically out of my room, but I don't think I was gone long.

B: I'm just wondering if you have any sense, now that you have just come back if you have any sense of whether you shortened the experience because you did not want to be there.

N: No. I think it was just short, very cut and dry. But, I think they implanted me with something and that's how they're going to get the information. They're not going to use my mind. Somebody is going to be in the area and get information through me somehow.

B: As you're sitting there on the bed now, is there any place in your body where it feels they have implanted something in you?

N: I feel like my arms were stretched outward very wide on the side of me when I was sitting in that chair. I couldn't see my hands, but it keeps coming back to me that my right hand and my bottom tooth in the right in the back.

B: And what is it about your tooth that you are sensing?

N: Something is in my tooth, under the crown.

B: Did you did have the experience of seeing other beings putting something in people's teeth?

N: I did, but that was during another experience last December, with different beings.

B: Is there anything else that you are aware of having just come back and sitting on your bed?

N: I think there is an unpleasant smell about these guys.

B: Just notice that smell, and get into it and describe the smell.

N: The smell might have something to do with a medical procedure.

B: Whatever you are resistant about, at least you're remembering it rather than having no memory at all. Any more thoughts or feelings as you're going to the bathroom and going back to bed?

N: No. I just want to come out of this now.

B: Okay. So, you're going back to bed and to sleep.

(End of Excerpt)

DECEMBER 28, 2005 INCIDENT

Journal Entry December 29, 2005

Last night, I had a very odd experience. I went to bed about 11:00 P.M. and fell asleep instantly with Murphy (dog) on the bed with me. About 1:00 A.M., he woke me up. He was behaving anxiously and making strange, whining noises. He was very frightened of something in the room that I couldn't see. Over the next half hour, nothing I did helped him and he refused to go outside. Still fully awake, I had just gotten back into bed when a blinding flash of white-yellow light appeared on the outside of my bedroom window that quickly moved into the room. I was very startled by the light, but instead of investigating, oddly I pulled the covers over my head and became unconscious instantly as though someone had flipped a switch.

I woke up abruptly at 3:10 A.M. Murphy was awake and still very anxious, moving back and forth on the bed. Again, I attempted to let him outside, but he backed away from the door and ran erratically between the sliding glass door and the front door, whining, for the next several minutes. In the two years that Murphy had been with me, never had I seen him behave in such a manner. Finally, I returned to bed and later awoke at 6:00 A.M. I looked for new markings on my body, but I couldn't find any. I still felt certain that something had happened that night.

Excerpt from Hypnotic Regression January 26, 2006

B: We are going to go back to the night of **December 28, 2005**, at eleven o'clock in the evening. It's time to go to bed and you're wrapping things up. Your dog, Murphy, is on the bed with you and he's very restless this particular night, more agitated than you have ever seen him before. He is moving around a lot on the bed and it's difficult for you to go to sleep because of his behavior. You're wondering what's making him so anxious and restless, and you're doing all you can to comfort him so you both can go to sleep. Now you're back in bed. Say aloud whatever you are aware of.

N: I hear a high-pitched sound with my left ear.

B: Does it seem like it's far away or close to you?

N: It seems like it's coming from outside and inside my head at the same time, but I'm hearing it with my left ear.

B: Does it seem like you've ever heard this before?

N: Oh, yes.

B: Is there anything, in particular, you're thinking about regarding that sound, or do you have a feeling about it?

N: I didn't realize that I had heard it because I was paying more attention to Murphy, to calm him down.

B: And is your dog still restless at this point?

N: Oh, yes.

B: Are you able to lie down now?

N: Yes. I'm hoping for the best.

B: What's going on now?

N: I'm cold and pulling the covers up over me, anxious to go to sleep because it's late.

B: Can you still hear that sound?

N: Yes. It's like there is a part of me that knows something is going on, but I'm not paying attention to it.

B: What does that say to you right now?

N: There's somebody outside.

B: Okay. Where might that be in relation to you?

N: They're to my left, outside of the back bedroom window, by the bushes.

B: What's happening now that you sense somebody is outside by the window?

N: I'm trying to pull the covers up over my shoulders and there's a big flash of light, bright light. It's white with a little yellow and it's so bright!

B: Can you locate the light?

N: It flashed two times. The first time, it was outside and then in another second or two, it flashed again and now it's inside my room. It' reminds me of the flash when you take a picture with a camera. I think they travel in the light waves! I just went out cold, but I don't feel like I'm out right now.

B: Some part of you is still there, right?

N: Yes.

B: Okay, with the part of you that alert, just go slowly.

N: I know there's more activity going on now, not just with the regular life forms. There's something about me that's attracting more attention.

B: Do you happen to know what that might be that you're doing?

N: I think, because my consciousness is focusing so much more on these matters, it arouses their attention. My mental focus creates some kind of energy that attracts them to me more. I don't know why I'm going to say this, but I feel like there are seven different types in my life now. Seven! This is the seventh one, and it's something different. They have a different kind of body and they travel differently. There are spots on their body and they're cold and clammy with weird joints.

B: Have you ever seen this before?

N: I don't think so and I've never seen this kind of light before, ever. I feel like they travel in balls of light that hit their target and boom! The light explodes and they come out. So weird!

B: So, when it hit your room it exploded?

N: Yes, and they came out of it.

B: Is it one being?

N: There's two or three. Murphy picked something up when they came. What's strange is that they have ears. None of them ever has ears.

B: So you can see them clearly enough to see that they have ears?

N: They do. These have pointy ears and their bodies move strangely. They're a humanoid form, but different.

B: What about their size?

N: They're a little taller, perhaps 5'10, but they're crooked with bigger heads and a leaner body with knobby joints. Their skin is clammy and cold and there are blue-grey and white splotches on it. They don't walk flat, but sort of levitating. They're around my bed now and leaning over the bed and me. They're not going to take me anywhere. It's as if they're just curious.

B: Now, when you say they moved sort of over the bed, does it mean that they're floating?

N: Yes, they float.

B: Are they looking at you? Are they aware that you can see them?

N: My covers have come off.

B: Does it seem like you would have taken those covers off or did they take them off?

N: They did. They're looking for anomalies within our race, scanning, and watching. My hands mean something.

B: Does it seem like they're particularly interested in your hands?

N: Yes, there's something about my hands that's significant to them.

B: Are they're doing anything particular, like touching your hands?

N: My hands are in a funny shape now as they're looking at them. I can feel it. My fingers are bent funny.

B: Looking at your hands right now, I can see that your ring finger is bent.

N: This is the first time one of these beings has not made me feel threatened, because they're just curious. I don't think they're doing anything to me, only looking and checking on me.

B: So they're not touching you?

N: No, they don't need to. They seem curious like a child would be curious. I think these guys are from far, far away.

B: Do you get the sense that they're three dimensional in size.

N: Well, I don't think they're heavy or substantial in their physical matter. They're more thin and fine and they even glow a little. I once saw a being that also glowed, but he was white with a blue tint under his skin. These are more like beings made of light. You can see the outline of their form, and the spotted, blue-grey color of their skin, but it's transparent. I think they reduce their matter to light waves.

B: That's very interesting.

N: If they can move their matter on light waves, then they can travel 186,000 miles per second like a light wave!

B: Oh, wow!

N: They can come from very far.

B: Yes.

N: And being more transparent, they wouldn't necessarily function on the physical plane the way we might think. I think that energy in the room knocked me out, not them directly.

B: With their light?

N: Yes. The covers are off and I'm cold. Murphy is beside me, but he's not moving anymore.

B: Does he seem to be sleeping?

N: He's awake, but just not moving. Now they're looking at my legs for some reason, and two of them seem to be communicating.

B: With each other?

N: Yes, they're memorizing all of this. I see also that they can touch each other to pass information, too.

B: When you say that they touch each other, what does that mean?

N: When they're side-by-side, their bodies touch, and information passes between them.

B: What do they seem to be doing now?

N: They're leaving now and they're on the other side of the bed. The light is flashing and they're just blending into it. Wow! It flashed bright again and went out of the window! It's gone!

B: Describe the flash of the light.

N: It had pointy spikes and it was glowing, and not round. When they were going, they just called the light back. Flash! It was there, they were in it, and then they were gone. I'm wide-awake now and sitting up. It's ten after three.

B: Did the light seem to go through the window?

N: The light went right through the window. They change their molecular structure to match the light waves and travel in them!

B: So, they're gone and now you are sitting up?

N: Yes, I'm in the bed sitting up, looking straight ahead and the clock says it's ten after three. Murphy is asleep. I always come back at 3:30 with the other ones, but I didn't go anywhere this time.

B: Did they do anything to you?

N: They just came to look at me, but they communicated differently between themselves. I might be able to pick up some extra information from them, not that they communicated to me. They didn't address me directly. My knees are so different than their knees. They kept looking at my knees and my legs.

B: While they were looking at you, was Murphy right there on the bed?

N: Murphy was numb, frozen, and not moving and then he went to sleep.

B: Okay, are you still sitting up now?

N: I got up at ten after three, got up, and walked around a little bit. I knew something happened.

B: Let's move forward now to waking up. This is a weekday, so is it time for you to go to work?

N: Yes, I'm getting up a half-hour late and I have to hurry. I have to leave in an hour. Murphy doesn't want to get up. He's out cold

B: Looking back on it now, how do you feel about what happened?

N: I'm surprised there are so many kinds and that they don't interact with each other. These don't have anything to do with the others.

B: With the other beings?

N: Yes, this was like a drive-by.

B: You mean like a drive-by in the sense that you don't have as much real contact?

N: A drive-by, in that they're not that focused on you. They just drive-by and stop, momentarily, out of curiosity. It's not like the others who are completely absorbed in us here. I think my sensitivity is increasing because of the contact that's been happening, and it makes me more receptive to other life forms, as well. Some part of me is changing.

B: Are these the kinds of beings whom you would be okay with if they came again?

N: I don't like any of them and I don't want any of this to be happening. I don't like surprises. I like to be asked.

B: And as far as we can tell, they're not asking you, are they? Any of them?

N: I've never been asked.

B: Do you find that you have some curiosity about them?

N: Yes, all of them. I want to know now what's going on. I like the truth.

B: We're going to be slowly making our way back from that experience. You're just coming back to this time and this place and this situation today and I will just be reminding you that this is Saturday, January 26 in the year 2006.

(End of Excerpt)

APRIL 8, 2006 INCIDENT

Journal Entry April 9, 2006

I had an abduction dream last night of alien invasion of the planet that was very involved. I was at my home, anticipating the arrival of my friend, Pamela. A woman from work, Debbie, someone I don't know very well, was also at my home. It was nighttime and I had walked out onto the porch. I looked up into the sky and saw a gold light, way up in the night sky and moving fast and zig-zagging, changing directions quickly. I called Debbie out and she saw it, too. I knew they were here and coming onto the planet. All of the electricity went dead in the house and the surrounding area. There were no lights anywhere in the house or the neighborhood. I thought of trying to call Pamela to tell her, but I couldn't find my cell phone in the dark. I was looking for it and a flashlight. I felt in shock and kept going back onto the front porch and looking up. Within a few minutes, there were dozens and dozens of ships up in the sky. They were cigar-shaped, very large, and still quite high in the night sky. It appeared as if a beam of light was coming from the front end of one of the ships that spread out about in front of it a quarter of the circumference of a circle. Hundreds of smaller ships had exited from the large ships, and they were literally filling the night sky, but still at a great distance away up in the atmosphere.

I felt very uncomfortable because the lights never came back on and nothing was working in my environment. It stayed dark for a while, perhaps a half-hour, and then Debbie and I heard what sounded like a large crowd of people coming through the streets and mumbling something repeatedly as if they were in a trance. There were hundreds of people coming our way and walking in the middle of the street. As I was standing by the front door, a round disk about six inches across came flying through the air and landed at my front door, literally bouncing off it, and then it fell to the ground. It looked like white paper almost, but I saw that it was metal and capable of flying through the air by itself. It had unusual, small black symbols written on it with a small nickel-sized hole cut into the center and a notch or two on the edge of the disk. I knew, somehow, that the disk held information about me and it came to my house to indicate to *someone* that I'm one of the abductees they've been working with. I didn't want to touch it, so I tried to kick it with my foot away from the front door and I went inside. The crowd was just about in front of my house and I was thinking to myself that they were going to gather us all up into one group and pick us up, but I didn't want to go. The crowd was finally directly in front of the house and someone from the group (are they human or alien?) came to the front door as if expecting me to come out, but I didn't so the crowd moved on. I realized at this point that the aliens could make themselves look like us so that we're not freaked out. They can affect our cerebral functioning to cause us to see what they want. There's a robotic, emotionless manner about them when they try to look like us. I'm kept thinking about Pamela, wanting to call her and warn her, but I couldn't find my cell phone. I knew then that they would overhear anything we were thinking or transmitting. Debbie was very upset and kept talking about getting back to her family. I was wondering where my dogs were because they seemed to have disappeared. I'm thinking they have been *turned off*.

Next, it was daylight and I was outside of my front door again, alone. I looked up at the sky and could still see dozens and dozens of ships spread all across the sky, now much closer down to the earth. I suddenly found myself away from my house and inside of what appeared to be a waiting room in a hospital or hospital-like corridor. There were a couple of people with me who I knew, or that I should know. I realized then that *they* are now all over our planet and disguised to look like us, even as children. As I was waiting there for something, I realized that I felt weightless somehow and I began to think to myself that my psychic powers had suddenly grown and that I could levitate and move in the air if I tried. I do try it and saw that it was almost like unlocking from gravity and I could float a little bit off the ground. (Have they done something to our gravitational force of the planet?) I moved through the corridor, able to raise myself off the floor and accelerate a bit forward. Suddenly, Murphy and Molly appeared at the door of the hallway where I was standing and they looked like they were waking up after a long time. I gave them some water, and something to eat, but they still seemed very dazed. Have all the animals been *turned off* during the invasion and now they're just waking up?

What are the aliens up to I'm wondering? I looked outside of the window of the building that I thought I was in and saw a city off in the distance, a city's downtown area with many very tall skyscrapers. I saw a large cigar-shaped ship suddenly come down from the sky and brush the top of one of the buildings, causing it to topple sideways and collapse. The ship did this to several buildings as I watch, mortified.

Next, I found myself walking through a very large, open area like a warehouse or inside an airport terminal. There were many *intake* spots set up where people were in lines and being checked in somehow. I was wandering through this large hangar and I realized that half of the *humans* that I saw were not humans, but aliens. You couldn't tell by just looking at them, but they weren't manipulating their faces or bodies like us, freely, and there was no emotion being emitted from them. They seemed to be talking quietly amongst themselves and moving about in groups of two or more. It reminded me of the two Men in Black that were at the MUFON meeting in August 2005.

Excerpt from Hypnotic Regression July 16, 2006

B: Today's date is July 16, 2006, and we're focusing on an experience that happened previously, going back to the night of April 8, 2006. Nadine, we're going to go back to your extremely vivid dream or actual experience of seeing many large UFO's in the sky overhead that were releasing smaller ships, and to whatever other details followed seeing those ships. To do that now we're going into a nice state of deep relaxation. What is your first awareness that something unusual is happening?

N: I hear a sound at the front door of my apartment, a knocking sound, or a clicking. Nothing has ever happened at my front door before like this. There's some kind of commotion going on outside and I have a strong sense that I'm supposed to go out there. I'm outside now, in front of my apartment, and standing barefoot on the cement. Trabuco Canyon drops way down just ten feet in front of my apartment building and I can hear something down there making a sound in the canyon. My neighbor lady is here, too. She's in a nightgown, without her glasses on. She usually wears very thick glasses, but she doesn't have them on now and she seems out of it. I'm feeling fuzzy, but I know there are ships up there.

B: Can you see any ships?

N: When I walk out on the sidewalk and look up, I can see an object in the sky. It's pretty low and hanging over the canyon at the back of my apartment. It's a matt silver color, not shiny. Oh, I'm in the ship now!

B:　So, you're not on the ground anymore?

N:　No, I'm on the ship and standing in a room with only females, female humans. There's a lot of us standing around in our nightgowns or nightclothes. Everyone seems pretty out of it with their arms hanging down at their sides. This experience has something to do with babies and war, taking care of babies at a time when there will be a big war. They're showing us smoke and bombs. Will they rescue earth babies at a time of war? The DNA has been altered in some of these earth babies, and they have added some of their own DNA into them, to change the human race and get rid of the aggression. Only some of the babies are going to be saved from the planet when the disaster comes. The extraterrestrials alter their DNA when they're inside the pregnant mother by injecting DNA into the womb of the mother with the baby inside. We are the protectors of these babies.

Oh, there's something wrong. Three races of extraterrestrials are here right now and one of them is a Reptilian race. I think there's something wrong with them and maybe they're dying. I can see different types of beings in different types of vehicles now and it's the Reptilians that have the cigar-shaped crafts. The little Greys and the big white ones with the big heads are here, too. Who is it that's dying?

B:　Can you ask these beings who will be dying?

N:　They're telling me that their cells are dying because they have exposed themselves to conditions here that are different from their own. They might have to mix their cells with human cells to survive here for very long because, right now, they can't stay here for very long at any one time. They take the blood from the babies who have been genetically altered, and they inject it into themselves so they can stay here physically for longer. They need some ingredients of our immune system to be okay here. They're telling me that this is a time of urgency now. Someone, a big white one with a big head, is very close to me now and looking right into my eyes to see the information I'm gathering. It feels like when they're looking through me they see inside my brain.

B:　Describe more how that feels to you.

N:　It makes the back of my head feel like a screen and he can look through my head, as if looking through a tunnel, into my mind to see the screen in the back of my mind. You know, I have always had this weird ability myself, to stand before the screen at the back of my head and write information on the screen. Then I can look at the screen later to remember the information I had written down. I've done this many times.

B: Does this experience of him looking into your eyes seem familiar to you?

N: Oh yes, very familiar. He's looking at stored information units, but they can't see or know my intention because it's not stored in the same way because they can't understand anything that's tied to emotion, and intention is tied to emotion. They can't grasp the concept of a loving God or love or compassion, in general, the way we can. They watch it happen in our minds, but they can't grasp it because there are different structures and receptors in our human brains that allow emotion and connection. The extraterrestrials don't have these structures so they can see only the objective information in my mind. Sometimes I think that if we weren't so scared when they look into our eyes like this, we could even see information about them in their eyes.

B: Are you afraid of the white one who's looking into your eyes now?

N: Well, not exactly anymore because my perception of them has changed over time. My favorite beings are the bugs that look like a grasshopper or a Praying Mantis. They do a lot of research and testing and they remind me of laboratory assistants. There are four different kinds of beings working together at all times. Man, I don't know where all these thoughts are coming from right now. They just come up so fast. Maybe they're putting these thoughts in my mind.

B: That's okay. Just stay with the experience. Is there anything else going on?

N: These beings are interested in the coil that I'm wearing around my neck as a piece of jewelry, the Q-Link. They know I wear it for more control, to keep them out, and they're not comfortable with it. They're just looking at it and one of them touched it briefly. I've been in this room before, and I know this whole experience is an exercise to see if we're ready for when the time comes. Things are going to escalate and they'll reveal themselves when the chaos comes so we'll be more receptive to them.

B: Can you see anything else? Are they showing you anything about that?

N: They showed me a future when they're landing and registering huge numbers of people, but it hasn't happened. It's only a future possibility. If it does happen, in the future I'll react more quickly and effectively because I've been previously exposed to it. I need to come back to earth now. I have mud on my feet. I must have walked down into the canyon with bare feet and now I'm up out of the canyon and I'm outside my apartment. It doesn't seem like I was gone very long, maybe just thirty minutes or so.

B: How are you coming back? How do they bring you back?

N: I've been getting back by floating onto the patio on a beam of light. The door back here isn't open. The dogs are asleep. I'm getting back into the apartment through the glass patio door somehow. Wow, I'm very thirsty and I'm drinking water from my hand from the bathroom tap.

(End of Excerpt)

Recorded Conversation Following Regression

N: It was so intense, but it wasn't all real. Some of it was just a projection they showed me. I've wondered if I've been genetically altered because I've always felt odd and strange here. Even as a teenager, I dreamed that I was on a planet with two moons. My dad also always talked about wanting to be taken by extraterrestrials.

B: Was he an experiencer?

N: In one experience, he was there, too. This thing about the DNA and the cells is very interesting. I never really pondered before about genetics or about extraterrestrials taking altered genetic material from a human baby and re-injecting it into their bodies to boost their immune system. Because of our bacteria and viruses, it would make sense that they would need to be immune to these things or they would never be able to survive here. Have you ever heard about somebody being pregnant with a baby and having genetic alterations done to that baby?

B: Yes, indeed, with at least two women clients. One female client was expecting twins and she was abducted during that pregnancy by the little Greys. There was a taller, female extraterrestrial who approached her on the table with an injection needle, and this tall female felt around my client's pregnant abdomen. She came to a certain place where she was touching one of the twins and not the other one and she very carefully injected something through the walls of the womb, presumably into that particular baby. When she was finished, she said that that baby was just the way they wanted him to be. When my client's babies were born, one of the babies had his eyes wide-open right after birth, and he looked right into the eyes of the mother and the father and the attending medical staff, startling them all. The other twin brother who was born at the same time, had his eyes closed and didn't look into anybody's eyes for quite a while, taking the usual time that newborns do to really focus. This particular twin was apparently the one who was affected by the injection, probably an alteration to his DNA. Years later, he turned out to be an experiencer and to be very positive and articulate about his extraterrestrial experiences. There was also one other woman I worked with who was able to recall in regression having had something done to the fetus inside of her body through her abdomen. So perhaps that was a genetic altering, as well.

4

24 CASE SUMMARIES

1. Mary, July 14, 1995

Under regression, Mary described being visited by numerous types of extraterrestrials from the time she was a child and continuing into adulthood. During most of her encounters, Mary recalled being transported from her home on a shaft of light that would appear through her bedroom window at night.

Although Mary's abductions had generally been conducted in secrecy, under regression she told of one unusual instance while visiting at her grandparents' ranch. In that circumstance, her grandmother had attempted to stop both Mary and her grandfather from being abducted, at the same time, by shooting at the aliens, possibly even wounding one. Later, her grandmother matter-of-factly discussed the issue of abduction with Mary, telling her that she could stop the abductions by mentally pushing in her mind in such a way that it would cause the extraterrestrials pain and thereby thwart the abduction. During subsequent abduction experiences, Mary attempted to use her grandmother's suggestions but was successful only if she was extremely frightened. Mary also remembered seeing her grandmother interact with an extraterrestrial on several occasions when she was four years old and again at age ten.

Mary described having encountered several different alien species over her lifetime. She recalled often being visited when she was a child by beings that she referred to as Little Whites, and another type she called the ant people, who would respond to her when she would call out in her mind. She thought of them more like angels, than alien beings, and felt comforted by their presence. Mary described another type of entity that looked like a Praying Mantis, with green-tinted skin. Its eyes were large and dark and it had no discernable nose.

Its hands were somewhat human-looking but had only three fingers and a thumb, and no fingernails. Although the being had a mouth that did not seem to open, somehow it appeared to smile. When it communicated, its skin seemed to lighten in color. She believed this type of entity was a doctor of sorts and she recalled meeting with him in a room that was filled with bright light.

Another type of being was very tall with a straight stature and biscuit-colored skin that would perform painful examinations on her. Its clothing looked like a Japanese kimono, and it communicated through its mind, even when not looking directly at her. After the examinations, these beings behaved more pleasantly toward her and even seemed to convey regret for the procedures. The worst of these examinations were carried out under their direction by very short and ugly extraterrestrials with grey, elephant-type skin. Mary believed that the Tall Whites do not value humans nor care for them and, during these encounters, she was unable to look straight into their faces. Instead, she would focus upon the front or backs of their bodies that seemed to glow eerily.

Mary believed that the extraterrestrials' nervous system is different from humans in that they possess an electric and magnetic field that can attract, amplify, and record other electrical impulses. She said that the extraterrestrials can only breathe oxygen for only a short time before they become sick, and they can breathe other gasses, as well. She observed that rather than breathing through their nostrils, they breathe through membranes, like gills, that open and close, and they can only remain on earth for up to six hours. She did not believe that they become invisible in front of humans, but they can change their form so that they can appear as a human or another animal.

Mary described having had examinations performed on her that included having needles inserted inside of her chest, stomach, and head. She also described having a small apparatus painfully placed into her mouth and ears that she thought was intended to repair or heal her. Other instruments were used in her genital area and, on one occasion when she found herself in a round room with reddish light, she believed they cut open her abdomen and aborted a fetus. A year before meeting Barbara, Mary also discovered scoop marks and bumps under the skin on her arms, legs, and shoulders that were red and swollen initially, but disappeared the next day.

As a child, Mary remembered being in some classroom settings, with as many as thirty-six other human children, where they were being taught by an entity she thought of as Mrs. Green. The room was divided into three sections and she was seated in the middle section. When the class was over, the children moved from the classroom by floating, as if in a flying dream. Mary recalled being taught how to heal herself, and she believed that, due to her contacts with extraterrestrials, she had been healed of migraines when she was ten years old. In a recurring dream of being in one of these classrooms, Mary sat before a yellow and green chalkboard where, through no apparent means, writing spontaneously appeared. Simultaneously, someone was imparting to her the thought that she would be learning about alien philosophy and psychology. During the familiar scene, she was also told that after more aliens arrived on earth it would be her task to speak to groups of humans about aliens to reassure them.

Mary's young son, James, was also an experiencer who seemed to enjoy the contacts, often calling for his space friends to come and take him in the light. She described an experience that happened one night when she and her son were sleeping over at her best friend's home. Her friend also had a young son about the same age as James, who, unknown to Mary at the time, was also experiencing extraterrestrial encounters. At approximately 2:00 A.M., Mary got up to feed and soothe her new baby. As she was walking about in her friend's home, she noticed a very bright light coming in through the second-floor windows. She dismissed this curiosity by telling herself the moon was exceptionally bright that night, and promptly went back to bed. The next morning her son very excitedly told her, privately, that for the first time he and his friend were taken together aboard a ship in the sky by their space friends. He further told his mother that they each had had numerous trips with the same space friends, individually, but that this was the first time they had had an adventure together. In privacy, his friend also confessed excitedly to his mother that he and James had just gone with their space friends, and Mary appreciated the validation of her son's experiences.

On another occasion, Mary and this same friend traveled to England, and one night while staying in a city hotel, Mary was awakened by an intensely bright light streaming in through their third-floor bedroom window. In the light, she saw her friend coming back into the room from the outside and crawling over the window sill. She jumped down from the sill to the floor with a very strange, emotionally expressionless face. Mary recalled thinking that it was extremely odd of her friend to come in through the window in a room that was so high above ground level, in the middle of the night. Quickly, she found herself in a deeply sleepy state and went right back into a sound sleep, completely forgetting the incident until years later. Concerning her abductions, Mary felt personally honored that she had been chosen for contact by the extraterrestrials.

2.   Brian, April 7, 1998

Brian met with Barbara several times in 1998 for regression work to unravel the many clues that he had regarding contact with alien beings. He remembered having had extraterrestrial experiences since the age of four and felt that most of his experiences were positive, especially those which had involved Nordic, human-looking beings. He was comforted by the fact that his roommate had experienced contact with Nordic beings, as well, and even Brian's sister had told him that she believed she had been abducted. The Nordics looked like healthy, attractive humans with blond hair and intense blue eyes. During his interactions with this species, they seemed to be infusing him with positive energy, strength, and psychic skills. They also performed physical examinations of his genital, urinary, and rectal areas, after which Brian would experience physical discomfort and symptoms.

Brian's regression work established that his abductions did begin in childhood. On one occasion he recalled being alone in the back of his family's car when a bright beam of light appeared overhead, emanating from a dark object. He attempted to hide on the floor of the car, but the beam of light found him and contacted him, and he heard a voice telling him he was coming to the tower. He was unable to move as the light transported him through the car and into the air where he was taken aboard a craft that was quiet and still. He was met by three or four small beings with slender frames, chalky, grey skin and hairless heads, which were larger than his head. Their eyes were large and round, their features quite small, and they did not appear to have a gender. They reminded him somewhat of the little Grey beings that he had remembered from previous experiences, but they were not as slim.

During this abduction, Brian was placed on a cold table and physically examined in a way that reminded him of the examinations he had experienced by his family doctor. During his examination by the aliens, they painfully inserted a long needle-like instrument with a bulbous end into his nose. He cried out in his mind, asking them why they were doing that to him. They responded, also telepathically, urging him to not be afraid and telling him that he was important and they would need to be able to locate him from time-to-time. He recalled how his mother had comforted him during doctor visits and comparing the experiences, he concluded that the alien's actions were okay.

In another regression, Brian described another childhood abduction when he found himself in a tall, round room about twelve feet in diameter that seemed like a tower. Other children his age were also present in the room, sitting in small, child-sized chairs. Also present in the room with the children were adult humans, who Brian believed were the parents of the children. The interactions between the parents and children were being very intently observed by non-human beings who he described as being the size of a six-year-old child with grey skin and large hairless heads. Their eyes were huge and black, their noses very small and they had only small slits for mouths.

In a subsequent regression back to Brian's childhood, he recalled being in a round room with other human children, engaged in mind games while sitting in a circle on the floor. Extraterrestrials were teaching them to move a ball with their minds by imagining the movement, with feeling and concentration. While describing the experience under regression, he remembered similar settings when he and the other children were taught to move first a piece of paper with their minds, graduating slowly to moving slightly heavier objects. Brian and the other children found the mind games exciting and they liked their strange teachers who seemed pleased by the children's efforts, and who told them they were specially chosen and the skills they learned would be important in the future.

As an adult, Brian also consciously remembered meeting with Reptilian beings who experimented upon him with needles, and conducted sexual procedures on him, obviously very interested in human sexuality. One of these abduction events also involved his roommate, Chuck, wherein they found themselves on an alien ship and in an operating room, each undergoing physical procedures of some type. After this abduction and on several other occasions as well, he noted strange markings on his body that could not be accounted for that included a triangle-shaped bruise on each thigh. He also recalled an occasion when he was on a ship and was shown a hybrid baby, half-human, and half-alien, and was told that the child was his baby.

Brian also described the experience of finding himself in a dark room with a domed ceiling. Two male beings were performing a procedure on him that included puncturing his back and neck. This created pain to his skull and brain that subsided quickly, leaving no apparent residual effects. After having experienced numerous physical and mental examinations and procedures, Brian realized that the extraterrestrials were studying all aspects of his physical and emotional makeup, including his spiritual understanding, his energy field, and his reproductive ability.

In another experience, Brian was placed alone in a lying position in a thick, clear jelly-like substance that felt like soft foam. During the experience, he was deeply and gratefully moved as he received telepathic impressions about reality and the nature of the universe, and the part he was playing in the total scheme of things.

3.   Jennifer, April 13, 1997

Jennifer had a consultation with Barbara and was regressed for the first time in her life. Because she was aware of strange experiences taking place in her life, she had been led to attend three MUFON meetings before the regression but had never been hypnotized. Since her first abduction, while living in a rural community as a child, Jennifer felt that she had been somehow tagged and continually tracked. Her attempts to discuss her strange experiences with family and friends had been met with disbelief and ridicule.

Jennifer described her first memory of seeing an alien being in 1975, during the daytime while she was home alone. She described the entity as being very tall with bluish-green skin, with a large head and pointed chin, and very black, slanted eyes. Its nose and ears were quite small and it wore some type of headdress with a collar. The entity communicated with her telepathically during the encounter, during which time she felt a pressure at the top of her head. Jennifer thought the being who appeared to be intellectually advanced, was urging her to go somewhere with him, but she was fearful and did not initially believe that she had accompanied him.

During a second occurrence one evening that same year, after observing a very bright light outside her bedroom window, she suddenly found herself outside and standing before a tall, unfamiliar alien being. She had an overwhelming feeling that he was somehow preventing her from looking at him directly, but from side glances, she was able to detect some things about him. His eyes were slanted, expressionless, and black, with high ridges over his brow, and his body was grey and hairless. She was very frightened by the being who appeared to be highly, mentally developed. He was communicating to her in such a manner that it felt as though he were inside of her skull. Suddenly she was inside a huge domed room that she believed was on a craft. Inside the craft, there did not appear to be nails or welding on the walls of the room without corners, and diffused lighting came from an unknown source. Many non-humans were moving about in the room and suddenly she could see the face of the being who had escorted her through the room. She was shocked to see that he looked like an insect. He rendered her physically unable to move and held her mind in a suspended state while telepathically telling her that they had arrived to spur us toward a greater evolution before our species would venture into space.

Jennifer recalled still another experience when she was being led through a tunnel by a different type of alien being that was tall and slender, wearing a military-type uniform. Physically he appeared flat and transparent as though he was being projected like a hologram, a measure she thought had been taken to prevent her from seeing his true physical state. During the experience, she was led by the being past row after row of large cages containing many different species of animals that she believed were not from earth. These specimens, she was told, had been collected from different planets to be studied and tested, along with those that had been taken from the earth. If relocation of life forms would become necessary in the future due to planetary changes, they were determining which species could survive on planets other than their place of origin.

Jennifer believed the extraterrestrials created amnesia in her after an abduction experience but believed that she could permit herself to remember. She also believed they are unaware of how intelligent humans are. Jennifer's abduction experiences have left her with a desire to learn more about animals and plants, particularly those plants that would be edible by many species.

4. Matthew, August 17, 1995

Matthew was consciously aware that he had experienced a series of events when he had lost time, during which he encountered several different species of extraterrestrials.

On one occasion, Matthew experienced missing time while sitting cross-legged on the floor in the living room of his home. His friend, John, was also present in the room and sitting on the couch. Matthew noted that unusual shifts in time seemed to be occurring when he glanced several times at his watch. Between each glance, although it seemed only a second had passed, the time reflected on the watch indicated that thirty minutes or more had elapsed. Although the television was not turned on, he noted that John was sitting on the couch staring at the television, as if in a trance. Matthew then became aware of a movement to his left and saw what appeared to be a female being in a lab coat. He was unable to look at the entity directly as he suddenly found himself paralyzed. He also sensed a change in air pressure in the room and heard air rushing behind him. The being touched behind his right ear, creating what felt like a physical response inside of his head. Still sitting and facing forward, unable to move, the being telepathically communicated with him, telling him that they would not take long. Although his visual perspective was limited, he could see that she was short with long thin arms. Her hands had only three fingers and a thumb and were without fingernails. Her skin was smooth, beige in color and cold to the touch, as it pressed hard on a muscle in his arm. Although he was very frightened, he was surprised to note that he was also quite physically relaxed. Matthew had been suffering from an ongoing problem with his knee that was quite painful, a fact that the being seemed aware of. Through touch, she somehow deliberately intensified the pain in his knee to the point that he was aware of tears streaming down his face. Matthew felt she was testing his endurance to pain, but she told him that the examination was intended to diagnose his physical condition and contribute to his healing.

During this encounter, a second entity appeared who behaved in a very detached and reserved manner, who retrieved pain medication from Matthew's bathroom and queried him telepathically as to its use. Eventually, three entities appeared in the room with Matthew, the third being an extraterrestrial he described as having grey-green skin and also wearing a white lab coat with pockets. This being communicated to Matthew that their examination of him would take three to four hours and he brandished an instrument that looked like a red, plastic toy gun. He coldly suggested, telepathically, to Matthew that they were considering killing his friend, John, who was still in the room; Matthew believed he was given that information simply so they might observe his emotional reaction. Suddenly, they were interrupted by an overhead light that appeared in the room, its intensity fluctuating, and he noted what sounded like radio frequencies in the background. Hurriedly, the beings departed, indicating they would return. Overhead the light grew brighter and the beings stepped into the light and disappeared. Strangely, Matthew could see that John was still sitting motionless in front of the television, staring forward.

On two unique occasions, Matthew experienced what he considered very positive and deeply emotionally moving encounters with beings that he referred to as the stick plant people or onion people. They were very tall and extremely thin, transparent beings that appeared in a cluster of twelve to fifteen beings surrounded by bright white light that made discerning their facial features more difficult. Still, he could see that they had large, soft, dark eyes, very tiny facial features, pleasant expressions and they emanated a feeling of love. They moved simultaneously in a wavering and undulating manner, with their full attention upon him. Telepathically, they expressed to Matthew that they were protecting him from the self-serving behavior of the other beings who had visited him and they would continue to do so. Whenever Matthew spoke of his interactions with these beings in Barbara's Experiencer Support Group, Matthew would express deep gratitude for the encounters who he believed were also helping him to see other-dimensional, human-looking beings, which he increasingly observed in his home, especially when reading or passing by a mirror.

5.   Cindy, May 3, 1995

Cindy began having extraterrestrial experiences at an early age and could often sense when they were coming, resulting many times in a panic attack. As an experience began, she would feel an energy shift and hear a low humming noise. Once the beings entered her home, she would note unusual energy patterns and sparkles of light filling the air, and her ears would begin to ring. During these encounters, she was aware that her radio would turn itself off. Cindy believed that she had had many contacts with several different types of beings, including one she referred to as Zetas, who were similar in appearance to Grey beings, but taller and more technically evolved. They told her that nuclear bombs had destroyed the surface of their planet and, as a result, their planet's population was forced to live underground for several thousand years. Cindy recalled these beings showing her holographic images of nuclear fallout while telling her that their large eyes and chalky whitish-grey skin had evolved because of living underground.

Under regression, Cindy described several other types of aliens. One was approximately ten feet tall with a bell pepper-shaped head, and large eyes. Another was a green insect looking creature with no apparent gender and two huge bug eyes, similar to a Praying Mantis, who behaved compassionately and appeared to be spiritually advanced. She also described another extraterrestrial that she thought of as being a quite ancient species, similar to the depiction of beings she had seen on temples on earth. She believed that the depictions on the temples she had seen were encoded with instruction on how to enter different dimensions.

Cindy also described having seen Men in Black, strange beings that look fairly human, but dress as if from an era decades ago. To her, they seemed self-serving and able to sense our fear, actually getting high from our negative energy.

During one encounter, Cindy recalled seeing beings coming into her room through the bedroom wall. They immersed her in a beam of light and transported her to the backyard of her home. During the abduction, her husband was sleeping beside her but remained in a deep sleep. Cindy believed these aliens were able to speed up the vibratory rate of her body's molecular structure, and theirs, to enable them to go through material walls. She recalled that while floating in this light, she did not feel completely solid, and that sensation remained with her for the day following the incident.

Cindy was provided considerable information by the beings that abducted her, including the fact that some extraterrestrials may be from the fourth dimension. She was told that our planet now exists in a three and a half dimension and that the human race is currently undergoing a metamorphosis. They predicted that the planet would suffer many great earthquakes, floods, and rampant diseases, causing the loss of life of large numbers of the human race, but many humans would be taken off-planet by the extraterrestrials during the chaos. They also shared with her the fact that power grids exist on the earth, as well as extraterrestrial bases, including one in Altadena, California.

The extraterrestrials relayed medical and genetic information to Cindy. They said that the human race was engineered as an experiment and that they had personally removed various chemicals from human cells to attempt replication of human emotions in their brain. Implants, they indicated, have been implanted into human beings for various purposes, including tracking and for brain stimulation to induce various thought patterns and emotions. In further describing the implants, she was told that the less advanced extraterrestrials tend to use crude implants constructed from a variety of metals, as opposed to the more advanced and benevolent extraterrestrials who use fiber-optic implants constructed from gold and quartz. Regarding human health, the ETs said that certain memories could promote disease, but deep muscle therapy could release those memories and thus prevent loss of health. On one occasion, several beings showed her pictures of the neural activity of her brain.

Cindy also believed that eggs were taken from her and were subsequently fertilized in vitro. She described a vivid dream that felt like a very real experience, wherein she saw a baby with white hair and eyes with a blue iris surrounded by white. She recalled being told that she was the mother of this baby and that she should hold it lovingly. They told her that because the baby was part human and part extraterrestrial, it required bonding from her as its mother. Those babies that are genetically purely extraterrestrial do not need to require emotional nurturing and human bonding.

Overall, Cindy maintained a positive response to her contact with extraterrestrials and especially appreciated the information they imparted to her. She believed that she had maintained long-term connections with the various beings, some of whom she felt related to her as one of their special humans.

6.   Erica, October 25, 1996

Consciously, Erica was aware of having had visitations with non-human entities since she was a child. Approximately seventeen years before her meeting with Barbara, while she was living in Nevada with her husband, one night a female being with very sparse hair, holding some type of equipment, appeared at their bedroom door. The woman stood still in the hallway while Erica observed the being, attempting to remain calm. As the woman walked towards her and reached out, she noted that her arms were quite long and appeared elastic. She also noted that the being was shorter than she was and that blue light was reflecting from behind her down the hallway. The being's manner suggested to Erica that she wanted to guide her toward the light and into the hallway, so she followed the entity. At the end of the hallway, the physical walls seemed to disappear and she became immersed and floating in the beautiful blue light.   Suddenly, she found herself hovering inside an egg-shaped room with brushed metal walls and a floor beneath her that looked like black rubber. A tall, oval opening in the wall appeared before her and she and the woman seemed to float through it, still in the blue light. Inside the next room, she saw flat-topped tables on bases, blank windows, and a railing along the walls. A scent like rubbing alcohol permeated the air and the blue light, pulsing now, still bathed her body, keeping her calm. She realized that her nightgown had been removed and they laid her down upon a cold metal table with her legs hanging over the side and her hands holding onto a small lip around the table. Another egg-shaped door opened in the side of the room and several beings came in. Frightened and claustrophobic, her breathing became difficult as a metallic machine was placed a few inches above her face. The machine projected a light that felt as though it permeated into her skull, causing pain. Although they used no apparent restraints, she was paralyzed and unable to move or call out.

A group of beings with putty-colored skin and an impersonal manner had gathered about her on the table as though she was a project they were working on. She noted that the beings had extremely large heads with two bulges on either side of their skull. Telepathically, they told her that they needed to remove something from her skull and then insert something else to replace it. Once the machine had been removed from over her face, she felt better, but oddly as though her head had expanded somehow.

After the procedure, the female who had brought her to the location helped her to get up from the table while all of the other beings left the area. She noted that she was still undressed as she left the room and physically walked through another doorway into a shiny, smaller room with a domed ceiling. The room was relatively dark although some indirect illumination seemed to be coming from the walls. Looking up, she noted that the ceiling looked like the night sky with stars. She was then set down into an L-shaped, metal chair that was cold and uncomfortable. Sitting before her was another entity that she described as a teacher. The teacher caused her to see pictures in her mind that he said were from the earth in the future year of 2010. The scenes were dirty brown and orange with crumbled buildings and a rust-colored sky. The being told her that the condition of the earth would begin to change in 2002 when people would turn on each other in the name of religion and god, and annihilate those who rebelled against various religious sects. They told her that related killing would also occur on the North American continent, in the United States, and that it was unlikely that it could be stopped. The being told Erica that the extraterrestrials want to reunify all of humankind and eliminate the hostility between various nationalities, races, and cultural groups, and thereby prevent humanity from irreparably destroying itself and the earth. She was told that several groups of extraterrestrial beings on the earth and off-planet are very concerned about humanity, having created us and studied us for millennia.

The day following this encounter, Erica's daughter told her that she had seen someone in her room the night before. Also, while bathing that following day, Erica discovered an unexplainable deep, scar on her left breast, as if a straight line incision had been medically performed on her. She also noted bruising on her legs that looked like handprints and small spots on her arms as though she had been pricked with needles. Similar markings had been reported to Barbara from numerous other contactees following an extraterrestrial encounter and Erica discovered that another woman from her monthly experiencer support group had also discovered an identical straight, deep cut over her right breast after an extraterrestrial encounter.

During all of her abductions, Erica noted that the extraterrestrials always made an effort to calm her and they communicated only through telepathy and physical gestures. Because of her encounters with extraterrestrials, Erica has developed an increasing amount of psychic ability and healing powers, which she uses frequently in her life.

7. Steven, February 7, 1996

When Steven met with Barbara, he was aware of having had four very vivid dreams about extraterrestrial beings and having seen several UFOs in the sky. As a young boy, he recalled that a being that seemed like an angel had come to him and his mother, telling her special things about him that he did not understand. He sensed that he was supposed to do something important concerning these unusual beings, yet it was not evident until years later by his career choice what that mission might be. His only clue came from the visions he had had at nine years of age when he saw himself speaking to large groups of people about something of critical importance that they needed to know.

In regression, Barbara and Steven explored his intense interest in extraterrestrial beings when he described one of his very vivid and realistic dreams. In that event, he found himself in an area of the desert, alone and floating above the ground in the warm air for what he believed was a first-time experience. In the suspended and floating state, he looked down upon the reddish earth and cactus plants and noted that he could manipulate his movement and direction. He saw no roads, vehicles or footprints, but did observe at a distance several other humans, dressed normally, who appeared to be floating in the air as well. One man in particular seemed to be enjoying the experience as much as Steven. In response to Barbara's question as to whether or not his body felt dense and physical, he stated that he could not feel his body, which suggested he might be having an out-of-body experience. She suggested he move toward the horizon, but was firm in that he was supposed to remain in that area. In that suspended and floating state, he began to receive information from beings whose presence he could sense, but not physically see. The information seemed to be streaming into his mind at an incredible rate, which he considered might be due to his then non-physical state. He gathered from the information he was receiving that he was to become a spokesperson for beings from other worlds, specifically to alleviate fear and to reassure humans that extraterrestrials intended to be helpful and not harmful. Their intention would also be to assist humanity regarding the massive earth changes that would be taking place. In response to Barbara's question as to whether or not he was referring to a particular group or groups that would be assisting, Steven was unaware as to what exact species would be helping. He did recall that he had been told that many species exist from elsewhere, some that create fear and harm, but that others may have a more altruistic intent.

In his life, Steven was experiencing a growing sense of responsibility to warn others of the dangerous changes that would befall the earth. He felt that those people living along the coastlines of the United States, Japan, Holland, and any island need to move inland to avoid earthquakes, tidal waves, storms, and flooding from the oceans rising. Deserts and higher mountains, he believed, would be safe, while the Midwestern areas of the United States would eventually be underwater. Many people would eventually be rescued from the planet by the extraterrestrials, while millions of others would die. Those who would be taken off would live on extraterrestrial ships and later be returned to the earth when conditions had improved. Steven believed many governments and persons from elite sections of societies are aware of this plan, but that it is unknown by the general masses. He believed that those in power were attempting to maintain control as long as possible while, at the same time, they have been preparing the higher ground for themselves in the United States and in other countries, as well. Steven was told that the earth changes would continue to occur until the year 2012, but he received no information beyond that point in time.

In response to the question as to when he began to receive this teaching, Steven stated that he now knows that he has been in contact with extraterrestrials all of his life and that he was beginning to receive information about the earth's changes many years ago, often in increments during his sleep. He was amazed and grateful that he had been provided with a tremendous amount of information from the extraterrestrials and was pleased with the contact and their treatment of him. He was also appreciative of his unusual abilities and experiences, such as his experience of floating above the desert. He remains very committed to fulfilling his mission as he has been instructed.

8.   Marcia, July 16, 1995

Marcia experienced contact with extraterrestrials for most of her life, beginning when she was about four years old. She always looked forward to them coming because she found the contact enjoyable. Having come from a large family in the Midwest, she believed some or possibly all of her family members had experienced abduction at some time or another. Her contacts involved beings she called Zetas and another type of Grey being that did not actually participate in abductions, but when they appeared, behaved kindly towards her. These beings she considered her special friends.

In several instances, when she was between four and six years old, she was taken into a huge craft that was out in one of her father's fields, and, during the abduction, she recalled seeing her father down below her on the ground. She experienced the sensation of leaving her body, floating through the air, and finding herself on a small craft from which she was later transferred to a larger craft by Zeta beings. In that experience, Marcia experienced a sense of loving connection with three of the extraterrestrials with large, hairless heads and huge eyes, who suggested that she was unlike other humans in that she was a part of their Zeta family. Marcia found that idea appealing.

In her regression back to this experience, she also described being in a classroom environment where she was instructed to draw pictures with unusual instruments that produced light, while the beings paid close attention to the functioning of her arms. They indicated there was something special about her arms and, to improve their functioning, the beings bound her wrists and inserted an instrument into each arm. Marcia said the procedure created a pleasant sensation, and they told her that her art would improve as a result. The beings also instructed her how to mentally control her arm movements so that she could draw the images that they telepathically transmitted to her. During that contact, she believed they also used an instrument to place an implant into the back of her head.

Over the years, she formed a special bond with one of the large-headed beings with huge, black eyes. This taller, thin female behaved more lovingly to her than her mother had, and Marcia welcomed her visits. This particular being seemed to be very concerned about the earth's children who suffered from autism, and she trained Marcia how to treat them. Marcia credited her contacts with this particular being for inspiring the dedication she developed to the care of autistic children. She avidly pursued reading books about the subject, and she spent time volunteering at a center for autistic children. With great success, she promoted the unusual methods of treatment that she had learned from her ET encounters.

Marcia described another particularly noteworthy encounter with the special being in which she was awakened in the middle of the night by a loud knocking sound on her front door. Answering the door, she found the same extraterrestrial female holding a human baby in her arms. Upon entering the house, she gave the baby to Marcia to hold and proceeded to give her telepathic instructions regarding the treatment of the autistic child. As the being was leaving with the baby, Marcia attempted to touch the extraterrestrial to help her to remember the visit more clearly, but her alien visitor refused to allow this. The very next evening, Marcia was awakened by the feeling of moderate pressure on her chest and when she reached for the source of the pressure, she discovered the physical presence of a long, thin, three-fingered hand touching her. Looking up she saw the familiar face of the female ET bathed in bright light and emanating unconditional love. To Marcia, her appearance brought reassurance that the unusual event the previous night had been real.

During Marcia's regression work with Barbara, she described seeing orbs during her contacts with extraterrestrials. She also recalled being in classroom environments where she was shown complicated pictures of planetary systems and was told that she would one day write about them when the timing was right.

9.   Louise, May 4, 2004

Louise had experienced multiple contacts with extraterrestrials that she referred to as Reptilians, with bumpy, mottled green skin that was soft to the touch. They also appeared to have tails. Barbara regressed Louise to an event that took place in 1992 wherein she found herself in a large, round room with strange, indirect yellow lighting for which she could discern no actual source. She remembered seeing gold, Reptilian eyes very close to her face and feeling that there were some other beings in the room with her, as well. Someone seemed to be controlling her, and she felt a pressure in her abdomen, neck, and head. Following the pressure, she had a glimpse of an unusual baby that looked like a reptile being who appeared to have been taken from her own body and slipped into a jelly-like substance. As she described this experience to Barbara, she became very emotional and wept, describing it as a wonderful experience. She held the unusual baby in the increasingly intense yellow light, enjoying the continuing emotional stimulation until the baby was suddenly gone. She wondered if the baby had resulted from a mix between her egg and the reproductive material of an extraterrestrial, although she had no memory of how conception had taken place. Louise felt exploited and disrespected by the beings who behaved as though they were superior to humans. In her interactions with the Reptilian beings, they tried to impress upon her that they intended to promote healing and not to harm. She recalled that during her encounters with them she often experienced paralysis and, at times, sexual arousal. The Reptilian beings also conveyed to her that she was connected to them and was very valuable to them. They explained to Louise that although they do have the capacity for emotion, including love, it is quite dissimilar to human love in that they experience feeling at a lower, more dense frequency than humans do.

10. Diane, August 31, 1999

Diane was regressed by Barbara on August 31, 1999, when she was twenty-three years old, and she had several subsequent regressions with Barbara later that same year. She had grown up in the Midwest and had often seen mysterious lights in the sky. She was aware of waking up in strange positions or in different rooms from where she had gone to sleep. Frequently, she heard a strange humming, rapping noises, and at times found marks and scars on her body for which there was no explanation as to how they had occurred. Sometimes, she was suddenly overcome with an irresistible urge to fall asleep instantly, on the spot, and she feared some kind of encounter might be taking place then. Although she was very resistant to the idea that she was being abducted by extraterrestrials, she set aside her fear of being hypnotized and decided to try regression to find answers about missing time events.

In her first regression, Diane relived an incident when she was ten years old and she found herself on a craft, along with another child. She was hiding on her hands and knees under what appeared to be a conference table. She could see many other people lying down on the floor, apparently asleep or unconscious. She described the extraterrestrials there as shorter than she was, with a slightly muscular build, and a blunt manner of thought and communication. She recalled seeing her father in the craft also, standing in a corner and telling her that she would be okay and that they wouldn't hurt her. In this regression, she recalled that on several other occasions she had seen white lights in the sky overhead while she was driving at night on country roads, and she felt that they somehow affected her. She was also aware that before an abduction, she would hear a humming sound, and a very bright light would appear. She would see movement within the light before the beings would appear from it.

Diane described an encounter with beings that occurred during the winter when she was fourteen or fifteen years old. Although it was cold outside during the abduction, she could not feel the cold weather as she was transported through the air. She saw houses and people down below her, as if from a long distance, but did not believe they could see her. There were six beings involved in this contact, all wearing long black or brown robes. Their heads appeared awkward, as though they had downs syndrome, and they seemed pleased to see her. She noted a strange noise in their presence, which she discovered was coming from the craft overhead. On the craft, she was taught to use screens and buttons on a round table. She realized that her father was also present during the encounter.

As an adult, Diane took several trips back to her hometown in the Midwest. On one of these visits, while she was staying at her parents' home, she recalled hearing a knock on the wall and a voice that told her to wake up, that it was time to go. She also described having had many direct encounters with an extraterrestrial being whom she referred to as Isaac. Isaac told her that in the future an asteroid would hit the earth and cause great devastation, with contamination and a shortage of food and water supplies. Isaac told her that there was a plan for her to assist when that time arrived. Diane described her contacts with extraterrestrials like visions and, although her experiences were benign, she wanted to protect herself from future contacts.

11. Simon, June 10, 1997

Simon was regressed in June of 1997, in front of an audience at a major UFO conference. In the regression, he relived an experience he had had forty years earlier while stationed in the Navy. On a weekend furlough, he and a Navy friend were hitchhiking from the naval base to their homes in the early hours of the morning. At approximately 1:00 A.M. they stopped at a truck stop for a cup of coffee, which was about a dozen miles from their destination. They sat drinking their coffee and joking with a waitress for about fifteen minutes and went back out to the road to find another ride. Another fifteen minutes seemed to pass and they saw what first appeared to be a truck moving towards them, with unusually large, bright lights. As it came closer, it rose slightly above them and they grew anxious because they realized that it was one large light. The next memory Simon had was when another car stopped on the road to pick them up. Both men felt disoriented and they were shocked to learn from the driver that it was 4:00 A.M., indicating that they had lost nearly three hours with no recollection of what had happened during that time.

Once in town, they joked about their strange experience and the fact that they had not had any alcohol.   After this experience, over the next few decades, Simon began to remember bits and pieces of disturbing images in his waking hours and dreams. In the regression with Barbara, Simon recalled that he and his friend had been drawn into the light of the object and deposited into a brightly lit room with a curved wall and ceiling. They were placed nude on tables, side-by-side, on what appeared to be examining tables which reminded him of medical tables. He noted that his friend was asleep or unconscious, but Simon struggled to remain awake and aware of what was going on. Standing about the table were very strange looking beings who were intent on inspecting his body. They were very tall and thin, looking amazingly like large Praying Mantis insects with huge black eyes, and long, bent arms. Their skin was white and they wore white robes of some kind.

Belying their fearsome appearance, the beings behaved gently in the way they curiously examined the frightened men. Afterward, they communicated to Simon that they had been aware of him for a long time, and they were training him in healing skills so that he might assist huge groups of humans in the future after large-scale disasters would occur on the earth. The beings gave him a powerful ability to creatively project his thought outwardly to manifest things and to physically heal masses of people at a distance. They told him his role in future years would also be to educate his fellows regarding the reality of the existence of other life forms, and their presence on the earth. At the time of this regression, he had been promoting conferences about extraterrestrial beings for years, without realizing that he, himself, had been in contact with the Mantis beings when he was eighteen.

12. Rita, January 11, 2001

Rita first consulted with Barbara on January 11, 2001, at the suggestion of her husband who thought she might be experiencing alien contact. As an experiencer himself, he had seen her on crafts during several of his encounters. Initially, Rita was strongly resistant to the thought, but gradually, she became open to exploring the possibility. In the beginning, she did believe that she had some memories that were screening something strange that she could not remember. As a child, she had been afraid of the closet in her bedroom and found it difficult to fall asleep for fear of someone coming from it. Her mother had shared with Rita that she, herself, had experienced some kind of alien contact, as well as a cousin in the family who had dreamed of holding a hybrid baby. Rita suffered from fibromyalgia syndrome and had a phobia about having her belly button touched. She had consciously decided not to have children but was uncertain what precipitated that decision.

Under regression, Rita recalled that in 1995, as she began to suffer from a sleep disorder, she would often sleep in the living room recliner, rather than the bedroom, where she was fearful to sleep. During that period, her husband had begun to complain that he felt he was experiencing extraterrestrial contact, and his admission encouraged Rita to confront her own experiences. On this particular occasion, she awoke to find herself lying on her back on a cold, metal surface, unable to move her body. She was naked, with a strange being soothingly holding her arm, while she underwent a gynecological examination. She noted artificial lighting above her in the soundless room, aware that she was not alone, but afraid to look about. Someone, communicating with her telepathically, told her they needed to take

115

something from her body and that she was making an invaluable contribution. She guessed they were taking eggs from her ovaries.

Another regression revealed an experience when Rita was riding in the back seat of an automobile when the radio and the car suddenly stopped, and the female driver passed out. She then found herself in a huge hanger with a dirt floor where human-looking beings, dressed in formfitting jumpsuits, were milling about. She noted that the experience carried sexual overtones, and she recalled a tall, pasty-white being doing something to her right arm.

Rita has had ongoing dreams of extraterrestrials and, when waking from these dreams, felt the memory carried a strong sense as if they had actually happened. In one experience, she described seeing a missile hit Los Angeles, California and she watched people dying from radiation. During this event, her sister was also present, and they both observed a huge spaceship and helicopters flying overhead. She also recalled being shown some kind of written plan for ETs coming to live on earth.

In another dream-like experience, Rita found herself in an institutional type setting, amongst a large group of people huddled together. A big, lizard-type entity with a long, thick tail walked upright among them. He wore no clothing and carried what looked like a clipboard. The group, in a trancelike state, was lead in a line into a larger room with rock formations along the side. They were shown enactments of violent scenes, including sexual relations between a human female and a Reptilian type of being. The scenes appeared quite realistic, as though they were happening in front of the group, yet Rita considered that they might have been looking at a hologram. As the group observed these scenes, a Reptilian being noted their emotional reactions from a distance, while writing on a clipboard.

Then Rita was taken, separately, into another room where she met a second Reptilian being with dark, rough skin and yellow eyes with vertical pupils. He urged her to look through an interior window into a room where a third Reptilian male forced a somewhat vulgar looking woman onto a medical table where he sliced open her abdomen with a talon from his finger, and she was rendered unconscious. The being who had brought Rita observed her reaction closely. The experience felt very real to her and she believed they orchestrated the scene for the sole purpose of observing her emotional response of shock and disgust.

In a subsequent meeting with Barbara on March 24, 2002, Rita said that both she and her husband were having multiple experiences of abduction, sometimes separately and sometimes together. She was greatly distressed over a new scar she had found on the inside of her right forearm, a straight scar one and a half inches long. She had no memory of how it had been created. The previous evening, while sleeping on the living room sofa, she found herself unexpectedly standing, barefoot, outside of her home in the wee hours of the morning, in fifty-degree weather. Feeling as though she were in a vacuum, she tried to collect her senses, realizing that there were no lights on in the surrounding neighborhood. Under regression, she recalled that before being displaced from her living room, five balls of white light, each approximately the size of a softball, appeared in the room, emanating energy and somehow beaconing to her. Emotionally detached, Rita became enveloped in the balls of light that moved her across the floor without her walking. She could feel her physical body growing lighter and less dense as she and the lights passed through the solid front door.

Once outside, she recognized her neighbor and two other persons about a dozen feet away, who were also being escorted by balls of light while looking upward toward another large yellow-white light in the sky. She realized it was not the moon. Quickly, she found herself in another location with someone communicating with her in her thoughts and showing her different planets in various stages of creation. A familiar being with huge, solid black eyes appeared next to her, communicating about creation. Then several other beings brought to Rita what may have been a hybrid baby girl, several months old, and told her that the baby was hers. It looked more human in appearance than not, but the baby's skin was thin, pale, and translucent, and her eyes were dark and almond-shaped with small eyelids. Her nose and mouth were small and she may have had tiny lips. The child's larger than normal head was hairless, and her body quite thin and frail.

Although the baby seemed unfamiliar to Rita, she was overcome with a feeling of love for the child and surprised that the baby seemed to respond to her. An image appeared in her mind of having given birth to the child. A bright light was surrounding Rita and the unusual baby and she sensed that other beings were observing her with the child. They informed her that they needed to remove tissue from one of the veins in her arm to inject into the baby to improve his health. Then they removed tissue from her forearm that resulted in a thin, straight scar. Coincidentally, her husband revealed to her that he, too, had an identical scar on his forearm, but he never regressed to its cause.

Rita met with Barbara again on April 16, 2002, to discuss the meaning of her experiences of frequently seeing 11:11 on the clocks around her. Under hypnotic regression, she recalled a new experience where she was led by a Reptilian entity to a destination where she met a second alien with unusually white skin who did not look entirely physical. This being showed Rita a picture or hologram of the future, of a vivid apocalyptic scene by an ocean, with sounds of the ocean and pictures moving in rapid succession of buildings being destroyed by huge explosions. They told her that she would survive and it would be her assignment to calm and escort a small group of people into an underground cave. She recognized some of the people in the group, including a female employee of hers. She recalled being told not to let the group be deceived but did not understand the meaning of this. Rita sensed somehow that going down into the cave would genetically alter her and those she would be leading, in that they would not be fully human any longer. In this regression, the symbol of a semi-circle appeared with a line crossing through it, and she vaguely recalled having seen it previously and she thought it was a significant symbol.

On April 29, 2002, Rita met with Barbara for further regression work. Having been feeling fatigued for some time, she had been diagnosed with a nonspecific autoimmune disease, possibly lupus or fibromyalgia. During that visit, she described having had ongoing thoughts about a new planet being discovered in the spring of 2008. She sensed that the beings who had shown her images of the earth disasters had also given her information about the new planet and trained her to lead people to safety.

13. Katherine, June 19, 1996

Katherine was regressed on June 19, 1996, and told Barbara that she believed she had multiple experiences of abduction by extraterrestrials and that they had placed implants into her body. She believed that before having ET contacts, she could feel sensations in her body of the implants being activated. When the extraterrestrials appeared, they arrived inside a beam of white light that would eventually engulf her body. Katherine frequently experienced a strong desire to go to Sedona, Arizona and to England, a desire that she believed came from the extraterrestrials, so that she might perform certain work in the vortex areas. She also believed they had inspired within her positive behavioral changes, including dietary changes, becoming a vegetarian, and implementing other health care measures. Because of the positive impact of her extraterrestrial contacts, Katherine preferred to consider these beings as her spirit guides.

Under regression, she recalled an incident of contact that took place while she was visiting with a friend in her home in Prescott, Arizona, when both she and her friend observed a ship over the house. She also recalled instances of seeing unusual beings move through walls and seeing huge ships hovering over the trees in San Fernando Valley, California. While observing these crafts, she noted that if she shifted her position, these ships would tend to disappear and then reappear in front of her, according to her position. Katherine believed that she had some sort of special connection to the beings in the ships and that they were very aware of her.

Katherine was regressed on April 11, 1996, and described an ET encounter that occurred while she was driving her car. During that experience, she noted a whistling sound in her right ear, and a great pain developed in the back of her head, behind her ear. In this regression, she discussed her contact with beings called Zetas, who they are, and how their consciousness works. She described having a friendly, male Zeta mentor communicate with her about the fact that other realities exist and function at a different energy frequency. He said that only a minor adjustment is necessary for a human or another being to shift into that realm. He told Katherine that although they found humans confusing, they longed to experience their emotions, but without the turmoil. They were particularly interested in the nature of human love and laughter, as they also experience love and laughter, but in a completely different manner than humans. They further impressed upon Katherine that not only do they lack emotional depth, but they also have a much greater restriction of opportunity, freedom, and individuality than humans have.

In one regression, Katherine described finding herself in a small round craft, standing at a table similar to a massage table and surrounded by several of the Zeta beings. They were performing some kind of healing work on an abducted human male who was lying unconscious on the table. Katherine recalled them meticulously instructing her on techniques for working on the energy system of the man, as well as techniques for healers to intermingle their energy and surrounding energy in the atmosphere with the energy of the man in need of healing. Although before her regressions, Katherine had been unaware of the training she received from the Zetas, in her waking life she had been actively involved in learning some methods of energy healing. Eventually, Katherine left her corporate management job and established a successful practice in energy healing work. Because of what she learned in her several regressions, she welcomes her abductions by the Zetas and enjoys a feeling of kinship with them, believing the lessons she has learned from them are invaluable to her mission.

14. Ken, January 14, 2003

In his meeting with Barbara, Ken expressed a belief that he has had abduction experiences throughout his life, beginning in childhood, always feeling as if he was being chased by something unknown. As a child, he recalled that his father displayed psychic abilities and that he and his parents would often watch paranormal television programs that Ken found extremely intriguing. He related one occasion when the whole family witnessed a UFO flying over their backyard during one afternoon.

In his adult life, he had had numerous, realistic dreams of traveling in unusual vehicles to strange locations, dissimilar looking from the earth, and encountering non-human entities. Continually, he experienced seeing repeating triple and quadruple numbers on clocks and in other locations that included 1:11, 2:22, 3:33, 4:44, and 11:11. Often, he would be awakened each night to see these numbers on his bedside clock, and he would be magnetically drawn to look at a clock or his watch at exactly the moment one of these numbers appeared. He also experienced a painful, stabbing pain in one ear each time he began to read material about alien beings. Ken came to feel that someone was trying to get his attention by these anomalous events. Coupled with his awareness of having had many dreams of interactions with alien beings, Ken decided to investigate these dreams further and contacted Barbara.

Over time, Barbara conducted many regressions with Ken. Early on, in one regression Ken discovered that his mother and father had also experienced abductions and, he recalled an event when he and his mother had been abducted together by three beings he referred to as ant people. In another event, he recalled standing on a shoreline and looking across a body of water at a demolished city, and the remains of toppled buildings and crushed cars strewn about. There, he stood along a ridge of land where a twelve-foot high gash had been cut away by some catastrophic event. Standing beside him was a small Grey alien being who appeared curious as to how he would react to seeing such a dismal scene. The being told him that various kinds of catastrophes would be happening on earth. He, and others of his kind, would remove from the planet those humans with whom they had been working. They inferred that he would be one of the people temporarily taken off of the planet.

In another regression, he found himself in his home during the daytime with the windows covered in heavy drapes because the sunlight during the day had grown so dangerously intense that everyone was forced to remain inside until the evening hours. The Grey being told him that the intense heat and sun on the earth was due to the huge solar flares from the sun. He said this scenario was shown to Ken as a forewarning of the future, as the scene of the destruction of the city had been. He wanted Ken to alert other humans to the future conditions on the planet and to instruct them to prepare for these enormous changes.

In a subsequent regression, Ken recalled a vivid experience in which he found himself leaving his home in the middle of the night, driving a white van, and picking up four people while traveling from California to a field near Caldwell, Ohio. He had been unconsciously directed to make the journey and to pick up the other individuals who had also been prepared for the event. Upon arriving at their destination, they joined a long line of automobiles approaching from several directions and headed toward a large UFO that was sitting in the field. When it came to their turn to leave their van and enter the craft, everyone in his group exhibited strong emotions about leaving their loved ones behind on earth. Extraterrestrial beings greeted them and ushered them to bench-like seats arranged in a tall pyramid shape, similar to bleachers at a ball game. Ken estimated there were approximately three hundred people present and, when all the occupants were on board, the craft left the ground. A being in charge addressed the group and explained that they had been in contact with everyone aboard for most of their lives, and they were being removed from earth because great catastrophes were about to occur.

The being further explained in Ken's seemingly realistic dream that the humans taken had been prepared for years and genetically altered to allow them to survive in the atmosphere to which they would be transported. In this experience, he recalled arriving at a location that he believed was not on earth and entering a large round building with a scalloped-shaped roof. It was located on an arid landscape where he presumed they would continue to live their lives. Here they could observe the destruction on earth from glass-like screens. Although he was glad to be saved, he was grieved that his wife was left behind.

In some of the experiences Ken recalled, he found himself with fairly human-looking people with large dark eyes, long dark hair, and a pronounced widow-peaked hairline. They reminded him of the ancient Egyptian depictions of gods with similar large dark eyes and dark hair. One of the females, wearing a beautiful, long cape, glowing with iridescent rows of color, appeared to take a special interest in him, instructing him to fly and perform other work on a spacecraft. Ken's wife, also an abductee, recalled encountering this particular widow-peaked female during some of her experiences.

In several of Ken's regressions, he was wearing unusual glasses that provided normal vision in the front lenses, but from the side lenses, he could see into other dimensions. He described being in a control room on a craft while wearing these glasses. He noticed a large, male Reptilian come into his peripheral view and suddenly appear fully in front of him as if he had stepped onto the craft from another dimension. Ken ordered the being to leave the control room and watched him fade in and out, inch by inch.

Ken regressed to one event when he found himself on what he believed to be another planet that was very dry and barren. There he saw a huge pyramid with smooth, straight sides and no visible steps. In the distance, he saw a large, blue glass tube that emerged from the ground and extended in a straight line for several hundred feet along the surface before disappearing back into the ground. The tube was approximately twenty feet in diameter with curved metal ribs, spaced at intervals, and large enough to contain a vehicle such as a train. A few years after having this regression he was intrigued to see NASA photographs of structures that looked like pyramids on the surface of Mars. Some of these photographs also showed what looked like large tubes that came out of the ground, ran along Mars' surface, and then entered back into the ground again. These images were identical to what he had seen in his regression.

Two particularly interesting conscious experiences happened to Ken while he was at home. In one experience, he was awakened while lying in bed. Although he was aware that his wife was sleeping in another room, he felt that someone was lying beside him in the bed. Lying on his side, he reached behind him and felt a small, thin body that he recognized as a female. He could feel a hand touching him with three long fingers and then he fell into a deep sleep. In the morning, he remembered the event with a vague memory of also having been shown an unusual looking Hybrid baby whom he was told was his child.

In another experience at home, Ken's wife awoke from sleep and left the bed to go to the bathroom, where she found the door closed and light shining from beneath it. She had noted when leaving the bed that her husband was not in the bedroom and assumed he was in the bathroom. When he did not come out after several minutes, she opened the door to find him unconscious on the floor with two small, Grey beings leaning over him. They turned to see her looking at them, and the next thing she realized, it was morning and she was back in bed. Although she recalled the event, her husband had no conscious recollection of his participation.

15. Nancy, February 9, 2003

Nancy met Barbara and was regressed on February 9, 2003. She had a strong interest in space and the possibility of extraterrestrial life existing, but also had a fear that she had been abducted her entire life. Nancy was engaged in learning alternative healing methods and she felt her mission was to assist in healing her fellow human beings.

Under regression, Nancy relived an experience of sitting in a chair in her home when she began hearing a buzzing sound in her head and feeling a physical sensation as if her body was growing lighter and might disappear. A small ball of light appeared in front of her, as a gripping pain began in her stomach and her breathing became difficult.

She sensed the presence of extraterrestrials who were telling her not to scream and, suddenly, she found herself floating in an iridescent bubble made of some kind of membrane in which she could breathe again. She noted a vibration and a humming sound in the background as she continued floating, waiting for something to happen.

Next, she realized that she was on a table and being observed by three beings with inordinately large, unblinking eyes, one in front of her and one on each side. They were causing energy to penetrate her body, measuring and manipulating her, and using instruments as well. She considered they might be healing her, for as they continued to weave various strands of light across her abdomen, she felt a release of pain and a surge of energy. They communicated without speaking and as they handled her abdomen, focusing on her belly button, she noted that their arms and fingers were long, thin, and silvery in color. Telepathically they indicated that they were going to treat her womb with something that would prepare the cells of the lining, in preparation for the future when they would implant her with another child. The procedure would provide some kind of protection for her and the baby. Years later, she would wonder whether her genius son had been implanted by them.

Nancy stated that the beings tried to convince her that they were her guardian angels, and although they seemed to be behaving in a caring manner, she did not fully trust what they said was true. The extraterrestrials told her that there were six dimensions of reality and that they had come into the third dimension from another plane, intending to impart information to the humans with whom they work. They informed her that she was one of their chosen people and they would be guiding her in a special mission on earth. As of her last meeting with Barbara, Nancy continued to feel she was being watched and that she could be abducted at any time.

16. Caroline, June 23, 1998

Caroline was regressed by Barbara back to an event that happened when she was eighteen years old while driving to Louisiana with her husband. She had long wondered about the event, sensing that something profound had happened to her on the trip. Under regression, she explained that she and her husband were traveling together in their car when, without cause or warning, the car's engine suddenly shut off and they came to an abrupt halt. She glanced at her husband who appeared to be unconscious. Just then, she saw a large, bright light in the sky that was moving in their direction. In a brief moment, it had engulfed the car, filling the interior with static electricity. She desperately tried to wake her husband by shaking and yelling but was not able to rouse him. A being then appeared in front of the car and she became aware that he had come for her and not for her husband, stirring within her a familiar feeling that she had experienced something similar on other previous occasions. The light dimmed and Caroline experienced a tingling sensation running through her body as she was lifted out of the car and levitated upwards, watching the car with her husband inside grow smaller below her. She rose in the air vertically, feeling as if a magnet was pulling her, while a buzzing sensation filled her head. She moved through the bottom of what seemed to be a craft, feeling curious as to how she was able to move through a solid object.

Once inside, she remained suspended in a room that was flooded with indirect lighting, noting that a rounded ceiling and walls were around her and a stainless steel table below her where she knew she would be lying soon. She found herself on that cold metal surface, surrounded by beings with large, bald heads who stood about five and a half feet tall and reminded her of doctors. Their mannerisms were not threatening and she believed they would soon release her. As they conducted their examination, they caused her to rise from the surface of the table and rotate in midair, using equipment that allowed them to view inside of her. The procedure was not painful, but it created a strange pressure inside of her. At one point, they used an apparatus with a suction cup at the end to remove something from inside her body. Although the procedure was not painful, it created a curious pressure inside of her.

These beings shared with Caroline their belief that the human race is far more primitive in comparison to the other intelligent beings in existence, and that they were attempting to assist the human species to evolve. They indicated that they vibrate at a much higher frequency than life forms on earth. They were treating Caroline and other abductees so that they might be able to survive the higher frequency that the earth will be receiving in the future. The beings assured her that they would continue to guide her and she floated rapidly downward and back into the car where she felt sleepy and confused.

It appeared as if her husband had simply pulled off the road and they had both gone to sleep, yet she knew that something profound had happened to her.

After her regression, Caroline said that although the memory registered within her as an intense dream, she believed that something physical had indeed happened to her.

17. Charles, March 12, 2003

For years, Charles sensed that he was not alone, and that unseen presences were often with him. He refrained from marriage out of fear that a wife might be negatively affected by his bizarre experiences. For years, he approached bedtime with great trepidation, often attempting to stay awake as long as possible when he sensed something disturbing might happen. Charles had three regressions with Barbara wherein he learned that he was experiencing contact with non-human entities.

One of Charles' experiences occurred very late one evening when tall, thin beings appeared in his bedroom. Their heads and bodies were long and thin and they appeared to glow. He was startled by their presence and he noted that the light they emitted seemed to calm him and minimize his fear. Surprisingly, he found himself rising from his bed, vertically, where he passed through the ceiling, into the air and up through the bottom of a craft. He found himself inside a round room with a yellow ceiling, which he continued to pass through until he landed in a darkened room. He quickly became accustomed to the dim lighting and he saw that four beings, looking similar to those who had abducted him, were in the room also. Gradually, other types of beings appeared in the room, including one brandishing a medical instrument. It had a ball with small protrusions at the end and it seemed to be used for scanning his body. He realized that this being and the group of onlookers were communicating without using their tiny mouths, obviously communicating through telepathy. Charles inquired as to the reason for the procedure and they impressed thoughts into his mind that they were checking on the condition of his health. Continuing their examination, several beings moved another instrument over him, from his head to his toes, which evoked a physically warm, glowing sensation throughout his body, and left him feeling wonderful overall. He noted that his hands were glowing brightly, radiating heat and energy, and the beings instructed him to touch many humans when he returned to earth. They further indicated that, from time-to-time, they would return to renew that energy because the people on earth would need it to help them deal with the major changes that would be occurring on the planet and within the human race in the future. Charles' response to the experience was quite positive and he felt honored to be of help to other races.

In Charles' second regression he found himself in a room that looked like a laboratory filled with large canisters containing various liquids and various mechanical devices. These included glowing red balls that moved through the air and that did not appear to be man-made. Charles realized that he and several other beings who looked somewhat human were floating in a viscous fluid, presumably receiving nutrients through their skin. Considering the more porous appearance of the extraterrestrials' skin, he guessed the process might be their normal manner of receiving nourishment. Although completely submerged in the liquid, he was surprised that he was so relaxed, free from fear, and able to breathe.

Charles described another event that occurred while stargazing, while on a camping trip in the desert with friends. Overhead they watched as an oval-shaped craft appeared and hovered overhead about fifty feet from where they were lying in their sleeping bags on the ground. He realized that his friends had gone to sleep, while he struggled to remain conscious. In a few moments, he found himself surrounded by several Grey beings. Without actually touching him, they vertically levitated him and his sleeping bag fell away. As he continued ascending rapidly along with the strange beings, far below him he could still see his friends sleeping on the ground. He and his companions entered through an opening in the floor of the large craft and, once inside, the ship began to move. Gathering greater and greater momentum, it felt as though the craft were moving at a tremendous speed through space. Through a window before him, he saw celestial bodies as they flashed by.

The beings placed Charles into an enormous translucent tube made of fibrous material and filled with blue light. Lying comfortably in the tube, Charles saw his life flash before him rapidly. He felt as if he were in a purely spiritual and blissful state without his body, similar to what he had experienced in meditation. He was removed from his enclosure and floated through a huge door, along a walkway within the craft that was long and oval-shaped with curved walls. The beings presented Charles with catastrophic scenes that included missiles being assembled and launched, and earthquakes destroying cities on earth. The scenes filled him with emotional and physical distress. He wondered which major cities he was seeing, and then realized they were showing him a worldwide event twenty years in the future. Although he seemed to be observing the scenes as if from overhead at a great distance, he believed it might be some kind of projection from within the ship. Suddenly, the pictures became black, with intermittent flashes of red, orange, and green that spread across the land and water on earth, and he was overcome with grief. The next scenes were glimpses of huge underground facilities where people were taking refuge. The beings conveyed to him that the destruction might truly happen, and he would not be able to avert it. They had given him an assignment that he was not consciously aware of but required that he watch the sky and look for messages.

After the regression, Charles described having had a recent vision of being in an underground facility on another planet. A large elevator was used to transport people, dressed in white, up to the dark surface from the huge underground chamber inside the planet. The facility appeared to be powered by a huge ball of energy and light. Charles concluded from all of his experiences that the extraterrestrials are actively concerned about the fate of the earth and humanity. Therefore, they are trying to warn and inspire some people to prevent the destruction that could take place. He felt that by learning more details of his anomalous events, he was benefiting from the experiences and was better able to integrate the events into his daily life.

18. Paula May 1, 1994

Paula was regressed as a volunteer in a training course that Barbara was teaching at a hypnosis school. This training focused on regression work regarding extraterrestrial encounters. Under regression, Paula described a particularly vivid dream she had that seemed entirely real. She was standing in her living room when a fairly human-looking female with long blond and piercing green eyes appeared. She was wearing a tight-fitting jumpsuit and a green helmet and was standing beside a strange apparatus. She communicated to Paula, telepathically, that she would be taking her on a trip. They both stepped into the unusual piece of equipment, which looked somewhat like a large inner tube, and then were lifted from the floor and through the ceiling of her home. Speaking aloud to the woman, she asked her where they were going and the woman replied, in her mind, not to speak aloud, but to communicate in her mind.

A netting of some type was placed about Paula's face that stopped her vision. The female indicated that it was for her protection from the different frequencies to which she would be subjected. Paula sensed that they were moving very fast, transporting to another destination, and a prickly sensation spread throughout her body, beginning at her feet and moving upwards. Paula had the impression that they were cleansing her before going further. The netting was removed from her face and she realized that she was in a large domed room on a ship out in space. Other than a table and chair, the yellow room had no windows or doors and was empty. The floor had a glowing mirrored shine, but no reflection. The air was dense and the lighting diffused.

Before her on the floor were several young boys. One in particular, who seemed familiar to her, was wearing a gown or kimono. Although she was told that he was eleven years old, she noted that he was playing in a manner like a toddler. The boy seemed very comfortable in her presence and looked at her knowingly in a way that stirred Paula with a feeling of love. She realized that he was quite pale and frail-looking, with thin hair, but his eyes were large and bright. He indicated that he wanted her to pick him up and she did so. Shortly thereafter, the female whom she had arrived with came to take her from the room. Paula became distressed at having to leave the boy to whom she felt deeply connected. Intuitively, she felt that the boys might be physically suffering and feeling pain, although they did not seem to display it. She was assured by the female that they were protected from feeling pain.

It also occurred to her that the boys in the room might be her Hybrid children who had been taken from her, that she should stay with them, or bring them back to earth. The woman informed her that although she would be able to see them again, she would not personally be able to raise them because the frequency of the boys was different from that of humans. However, in the spacecraft environment, they could commingle. She further told Paula that her eggs had been combined with the reproductive material from members of another race to create healthier offspring, as part of a crucial project involving twenty-four male babies. She said that the human mothers had consented to participate and, as the boys grew, they would be asked to contribute to the reproductive project to save that race of beings from extinction.

After the regression, Paula wondered whether her son at home might have been contacted or altered by extraterrestrials without being aware of it in his conscious state, just as she had been unaware. Her memory began to release more information and she realized that she had been frequently experiencing nighttime visits by the extraterrestrials. She recalled being told that she is one of many humans assisting them and that humans are living in several dimensions simultaneously.

Paula felt supported and blessed by the beings and noted a vibratory response in her body whenever she spoke about them. She was comfortable with her discovery and expressed excitement at being involved in the reproductive project of the beings. Paula strongly felt that what she experienced under regression was reliving an actual reality that she could fully sense and touch.

19. Shirley, September 30, 1996

For all of Shirley's life, she had felt an uneasy sense of not belonging on earth. Driven by a deep longing to reunite with something or someone out in space, she frequently found herself obsessed with searching the sky. Although she was happy with her life and family, Shirley felt as if she had originated elsewhere and had been displaced. In her regression with Barbara, Shirley searched to find the source and the meaning of her strong sense that she belonged with other beings and not on the earth.

Under regression, she described feeling peaceful and seeing beautiful swirling colors that, within a few moments, gave way to a feeling of floating and rising upward at great speed. She found herself on an unearthly terrain and saw before her an enormous metal building with huge windows with disk-shaped crafts moving in and out from the structure. No people were present and the next thing she recalled was being inside the building, feeling a sense of emotional peace and serenity. As she floated along a walkway, several human-looking beings about six feet tall appeared in front of her, wearing maroon jumpsuits. They noticed her, but continued on their way without communicating. Shirley could see that the walls of the walkway were not entirely solid, and an opening at the end suggested that she was seeing daylight outside. She moved into a room where beings were present like those she had passed in the hallway. Although they seemed surprised at her presence, they greeted her warmly. There, she was left alone to view on a large screen a series of violent scenes of war that appeared to be taking place on earth. She believed the beings were impressing upon her their perspective of earth's future.

As the event continued, Shirley was aware of being in another, much larger room that looked like an airport terminal and she noticed an elevator was in the room also. Many people were moving about wearing the same one-piece jumpsuit that she had seen previously, both black and maroon in color. The beings looked relatively human, with brunette or blond hair, but they shared strikingly similar facial features and large intense eyes. Many disk-shaped crafts were coming and going from the terminal through openings in the building without making any sound. A huge picture of the earth was mounted along one side of the terminal. It was marked with handprints as if it had often been used for reference. A male being who appeared to be in charge acknowledged her and told her she was important, that through her they were able to learn more about life on earth and the dynamics of the human race. He placed his hand upon her forehead for several moments without speaking, stirring a pleasant emotional sensation within her.

After the regression, Shirley was very pleased to have been able to relive the contact with the beings, and she expressed relief at finding the cause of her ongoing sense of deep longing to be elsewhere. Overall, she viewed the experience as very positive and looked forward to further contact with these beings.

20. Madison, April 17, 1998

Madison was aware of having had many extraterrestrial experiences before doing regression work with Barbara, although she consciously remembered only fragments from each. In 1991, she recalled having had fourteen experiences that she was aware of and that the visits were mostly positive. Her original fear had given way to a positive feeling and a sense of love for the beings.

In regression, she described an experience from 1990. While sleeping at home, she was awakened by incessant barking from her dog. Upon rising, she experienced an electrical tingling sensation throughout her body that caused her hair to rise up from her head. She realized the whole house was filled with static electricity. Fearful, but intrigued, she faintly recalled having had the experience before and the thought occurred to her that they had arrived. She thought about waking her brother up so that he could witness what was happening, but she refrained from doing so because she felt that she was somehow not supposed to. Opening the front door of the house, she stepped outside and saw that a large craft was hovering nearby. Colored lights were pulsating on the underside and she could also feel an energy pulsating in her body.

Stepping back inside the house, Madison found herself walking through the hallway with three beings beside her, a strange green iridescent light filling the rooms. Passing her brother's bedroom, she saw that he remained in a deep sleep, unaware of what was taking place. She calmed her fear by recalling the fact that she was more afraid of some humans than of the beings in her presence.

One of the beings, who appeared to be the leader, stepped forward and began to communicate to Madison through his mind. He projected various pictures to her, including an interaction that suggested she had with them when she was seven years old. He also projected pictures of symbols and indicated that one of them, a hexagon, was used in conjunction with space travel. As the contact continued, the electromagnetic energy around her body increased significantly and she could see a blue ray of light around her. Her body felt strangely cold and she sensed they were preparing her for an exchange of information.

He telepathically told her that pictures would be transmitted to her in the future and that she did not need to understand their meaning. Over time, it would become clear to her. He suggested to Madison that other humans, such as scientists, were also receiving similar transmissions, and using the information in their work. He further suggested that she would be given time to incorporate her newfound awareness into her daily life before she would receive any new information from them. Then she would be able to share the information regarding alien technology and space travel with other people through speaking and drawing. Madison, fascinated by the prospect, felt a sense of kinship with the beings before her. At 4:18 A.M., she suddenly awoke in her bed with an imperative need to write down all she remembered, including the strange symbols, as she had done many times before.

Eventually, Madison realized that the symbols she had been drawing for years were the same symbols that had been showing up in crop circles in England. Therefore, she developed a great interest in the crop circle phenomenon. She wondered if the same beings who had been telepathically sending her symbols were the same beings that were making the crop circles in fields on earth.

21. Harold, August 27, 1995

Harold had many bizarre dreams of unusual beings and he experienced periods of lost time, but he was unaware of having been abducted until he was in his late forties. On one occasion, when he saw a model of a disk-shaped craft in a hobby store window, brief memories of a craft, which he called the sport model, arose to the forefront of his consciousness. He impulsively purchased the assembled model and took it apart, looking for a blue column or beam that he believed would be in the center. He did not understand why he felt so certain that it would be there, but a strong sense overcame him that he had seen such a ship in real life. A few years later, he heard a radio interview with Barbara Lamb discussing her regression work with people who were allegedly having encounters with extraterrestrial beings. He learned that she worked in a nearby town and he later contacted her and experienced a series of regressions with her.

In regression, Harold realized that his extraterrestrial contacts had begun while he was in his mother's womb. While his mother was asleep one night, she was visited by a Grey male being who placed his hands on her abdomen, in the area of her womb where Harold was developing. The being said directly to him that his heart was defective, in that there was a hole in one chamber that would make it impossible for him to live after he was born. The being also indicated that he was placing an invisible, electromagnetic strip over the defect in Harold's heart so that he would survive. The electromagnetic strip would be reinforced by him after Harold was born to keep his heart working properly. During this regression, Harold also recalled being taken from his bed at the age of three and being placed on an examining table for a physical examination by the same male being who had visited his mother while he was in her womb. On this occasion, he placed his hands over the little boy's chest, apparently reinforcing the electromagnetic strip. Over time, Harold grew quite fond of this being who also taught him how to move objects with his mind. Harold described him as approximately five-foot-seven or eight inches tall, slender with grey skin and large, faceted eyes. He referred to the being as Mazu. During subsequent regressions, Harold found himself on the same craft at different ages learning unusual skills. During these visits, the same action was taken to reinforce the strip over his heart.

When he was a young boy, Maze frequently escorted him through the ship and he eventually allowed Harold to explore the three levels of the ship on his own. In doing so, he discovered a beam of bright blue light that ran vertically through the center of the craft, from top to bottom. It seemed to originate from a power source on the lowest level. Repeatedly, he was instructed never to touch the blue beam that was used as a source of power and guidance for the craft, or he would be severely injured. Harold observed a group of Grey beings standing around the beam, staring into it unwaveringly with total concentration. This action seemed to control the movements of the craft.

When Harold was in his late teen years and early twenties he practiced standing by the beam and focusing his mind into it while the Grey beings stood nearby observing him. He would attempt to discipline his mind to focus on the movement of the ship, feeling their displeasure if he became distracted by other thoughts. In recalling this experience, Harold realized that the wide range of thoughts and feelings of human beings made the task of psychically directing the beam far more difficult. During each visit, after Mazu reinforced the electromagnetic strip over his heart, he continued the practice of disciplining his mind.

When Harold was forty-eight years old, Mazu instructed him to go to a heart surgeon on earth for surgical intervention. He underwent heart surgery with his wife, a registered nurse, observing the procedure. She observed the shock and confusion of the doctors, upon viewing the silver-dollar-sized hole in his heart. They estimated it had been there for many years, but for some inexplicable reason, it leaked very little blood. At the time of the surgery, the entire medical staff was amazed that Harold was alive. Both he and his wife were unaware at the time that extraterrestrials had been reinforcing his heart for years.

In his last regression with Barbara, Harold, who was fifty-two years old, recalled successfully piloting the ship by perfectly focusing his mind on the blue beam, while telepathically in unison with six other beings on six other ships. All of the beings, including Harold, were so telepathically connected that all seven ships flew in complete unison, directed only by Harold's thought.

22. Jessica, October 29, 2000

Jessica was aware that several unusual occurrences had taken place in her life, beginning when she was six years old. She and her family, as well as many other people in town, observed a large UFO, with lights on the underside, flying overhead in the evening sky. She also recalled that her father would often bring up the subject of UFOs in conversation, remarking satirically that her brother was not of the earth. As curious as she was about his comments, her father was never willing to discuss the matter further.

While traveling with a friend in Sedona, she and her friend observed a large ball of fire moving across the sky until it faded from sight. Later, during that same day, they observed a similar red object, at a greater distance, that moved in a very dramatic, erratic manner. They were both certain that neither object they observed was a conventional craft. A few years later, after her friend confessed that she believed she had been abducted, Jessica also began to question events in her life. She seriously considered that she, too, might have been experiencing contact, so she decided to undergo hypnotic regression.

In beginning her work with Barbara, the direction given to her subconscious mind was to reveal any important events involving extraterrestrials during childhood that would be important for her to know about. In regression, she found herself in a large metal room with diffused lighting that appeared to be a sterile environment like a laboratory. Metal tables with equipment overhead filled the room and she realized that she was about six years old. She was sitting on one of the tables, without clothing, and dangling her legs from the side. As she looked around the room, she saw that other children and adults were lying upon the metal examination tables, but they appeared to be more dazed and unaware than of their surroundings than she was. A human-looking female with brown hair, dressed in white clothing and a cap, and who looked somewhat like a nurse, came to stand beside the table where Jessica was laying, and she began to adjust equipment in the area. As a group of five thin, Grey beings entered the room, the female reassured Jessica that she would be fine.

The Grey beings surrounded the examining table and gazed down upon her. She saw that they were barely taller than she was with the outstanding feature of having huge, black eyes with no white surrounding area. They were hairless, without ears, and had only tiny openings for a nose. There was a small slit for a mouth that did not appear to open. The skin on their thin arms was quite rough and their hands had only three fingers and a thumb. The tips of their fingers were quite rounded and there were no fingernails. Behind the Grey beings who were standing around her, she saw another type of being observing her. He was much larger than a human and she referred to him as Godzilla. This lizard-like creature was darker in color and had bumpy skin and yellow eyes with vertical pupils. As she lay on the table, fully awake, an instrument of some kind was inserted into her mouth and down her throat. Soon, the others began to perform examinations on her ears, eyes, and genitals. Jessica consoled herself by holding onto one of her stuffed animals, which she had brought with her.

During the examination, she did not notice any communication between them, but intuitively, she felt they were checking her progress to understand something more about humans. Jessica learned that Grey beings are a dying race, without reproductive capabilities or emotions. When she asked them why they don't care about humans, they telepathically responded that they do what they have to do. They told her that she could choose whether to participate or not. She responded in her mind that she did not want to be involved. In asking them why they picked her, they told her that because of her father, she had the genetics that was of interest to them. Jessica eventually came to believe that her entire family has experienced abduction, including her brother, cousins, and grandfather.

At the end of the examination, Jessica was taken to another area of the craft where she met other younger, human children who were playing with toys. There were no sounds in the room and it seemed as if everyone was moving in slow motion as they played. Sitting on the floor, fully clothed now, she held onto her stuffed animal, noting that none of the children were interacting with each other. This room was also constructed of a silver-type metal, even the floor, and the atmosphere in the room seemed strangely humid and hazy. As she glanced about, she saw what looked like a human female with long blond hair who was watching the children and playing with a child. Jessica found her appealing and was comforted by her presence.

The next scene Jessica recalled took place when she was older. She found herself in a room with a Grey being and sitting in a straight-back metal chair while the being attached an odd metal brace with wires around her head. She could also see what looked like other humans in the room. The Grey being placed something on her skin and into her mouth, creating an electric current that moved painfully within her head. The pain caused her to bite down onto a wet cloth-like material that had been provided to her. As the electric current moved through the rest of her body, escalating the pain, several male humans and male Grey beings observed her dispassionately and communicated together, without her understanding what was going on. Because of some of her experiences, Jessica felt fearful at night, especially when getting up to go to the bathroom. Some nights, both she and her roommate have felt a strange energy in the house.

Jessica told Barbara that she had experienced contact with other, more loving, beings who appear sometimes when she calls out to them. She believes that they are teaching her and providing her with important information in her sleep. They have also told her she would no longer be abducted by the Greys. By the end of her session with Barbara, Jessica felt she was restored to her good energies and could resume her work as a massage therapist assisting others.

23. Alice, September 5, 1999

For several years, Alice was aware that her husband had been experiencing the strange phenomenon in his life that appeared to be in contact with extraterrestrials. On one particular occasion, she decided to accompany him on his appointment to see Barbara Lamb. As Barbara guided her husband into deep relaxation to access the memories of an experience he wanted to explore, Alice also went into a deep altered state of consciousness. In the hypnotic state, unusual symbols appeared before Alice in her mind, and she experienced flashbacks of her own experiences with unusual beings. As her husband's regression progressed, she drew on paper what appeared to be an unknown language. The experience affected her so deeply, that she subsequently decided to pursue her hypnotic regression work with Barbara.

When Alice began her work with Barbara, she did come to find that she had been experiencing contact with extraterrestrials. The first and most important event that Alice wanted to explore was a period in her life when she had experienced three days of missing time with no memory of what had occurred. In that regression, Alice found herself being taken from her bed at night by three short beings with white, hooded robes who floated her horizontally up through the wall of her room and into a bright beam of light. Once in the beam with the other beings, they moved upwards into the bottom of a large object hovering in the air.

Inside, Alice was ushered into a plain, beige-colored room where more of the robed beings were waiting beside five large, spherical containers, approximately eight feet in diameter. Each sphere was lined up in a row and filled with fluid. Her clothing was removed by the beings and she was floated upward and placed into one of the containers through an opening in the top of the container. Once inside, she became immersed in the fluid that was thicker and more viscous than water but allowed her to breathe without difficulty. It seemed as though she remained in the container, which she referred to as a pod, for up to three or more days, while she imagined the craft was traveling a long distance. In the pod, she remained comfortably floating, breathing easily, and moving in and out of conscious awareness. The pod continuously rotated soothingly like a gyroscope, creating a blissful experience for Alice that seemed vaguely familiar. A porthole on one side of the container allowed her to peer out occasionally. Sometimes, she would see the face of an alien-looking back at her. He appeared to be concerned for her welfare and to be checking on her condition.

Eventually, she was removed from the container, dripping with fluid and coughing intensely to expel it from her lungs so that she could breathe air again. Two of the beings in white, hooded robes assisted her with drying and dressed her in an identical robe. As they were doing this, she noticed five other humans were being tended to similarly. They were standing next to the pods they had traveled in. One of these persons was having a very difficult time trying to expel the liquid from his lungs and finally collapsed on the floor, unable to breathe. Because she was a nurse by profession, she attempted to go to the man but was restrained by the beings. They impressed upon her the urgency to move forward without him, and to move on to an important destination and to arrive at the proper time.

Alice and the four other humans were led, single-file, along a winding, beige-colored corridor. As they moved through the corridor, each human was escorted by two beings, one on each side. She noted that these robed beings were about five feet tall, but she was unable to see their faces as they were hidden by the stiff hoods. They approached a large oval opening at the end of the corridor and entered into a huge room that reminded her of an indoor sports arena. It was brightly lit with many tiers curving around the room and wide steps leading to the top row. Beings, dressed in identical hooded, white robes sat side-by-side on each tier before connecting tables. The scene reminded Alice of the General Assembly of the United Nations. In the middle of the main floor was a long table where three robed beings sat, with the center figure wearing the only different costume in the room, which was a blue robe. Alice could not see the faces of anyone present. Due to the different shapes and statures of the figures, she realized that many different types of beings were present. The hooded robes concealed the different species, thus preventing distraction. Alice was fascinated by the gathering and wondered if her group was intended to represent the earth.

The blue-robed leader telepathically projected a greeting to those at the gathering, allowing her and the others to understand in their language. Various issues regarding the different races and major changes in the universe were topics of discussion. Planets and star systems were mentioned that were unfamiliar to Alice and referred to by what seemed to be energy or vibratory frequencies. Earth was a topic of discussion with a focus on pollution, destruction of oceanic life forms, and the sonar frequencies that enhance the planet. The danger of humanity's nuclear capability and its potential impact on the solar system was a concern addressed. Another topic of discussion was the negative characteristics of humanity, especially our competitive and warlike nature. They said that a number of the races present had been working for centuries with human beings to promote cooperation, love, and compassion. The leader charged the humans present in Alice's group with the responsibility of organizing large-scale activity on earth to bring about change and asked others in the assembly to take further action to influence human genetics to eliminate the dysfunctional aspects of humans. The meeting concluded and Alice and the other four humans were ushered from the room, returned to their pods of liquid, and sent home.

When Alice awoke in her home, she realized that she had been gone for three days and she had no memory of what had happened to her. Her husband, who was away on business, had not been aware that she had been missing. Before her regression work, Alice had not been able to recall any memory of those three days and had dismissed her lack of memory to the fact that she had been going through a stressful period and was exhausted.

24. Sara, March 9, 2004

From her earliest memories, Sara had been aware of having extraterrestrial visits. She shared a bedroom with her younger sister and very often, they would both see shimmering white lights coming in through their bedroom window. The lights would create patterns in the air that they referred to as fairies. Eventually, the lights would settle to the floor and become three short white beings with large heads, spindly bodies and limbs, and huge round, black eyes. They appeared to have only small openings for a nose, tiny mouths and they glowed in the darkened room. Sara and her sister were comfortable with the presence of the beings who appeared to be gentle, and they enjoyed spending time with them. After they arrived, another bright shaft of light would appear and shine directly on Sara and she would be absorbed into the light and floated from the room. Her sister would be left behind, in a deep state of sleep. Through her regression work, Sara learned that she had had many experiences, well into her middle years, with the same beings who would take her away with the beam of light through a wall, closed window, or ceiling. She considered the beings gentle and sweet, and she referred to them as her little white guys. Although these beings were the only ones who removed her from her home, during the ET experiences she was taken to many unusual settings and interacted with a large variety of extraterrestrial species.

In one experience, when she was in her late thirties and was married with young twin sons, she was taken by the Small White beings to a room on a craft and seated in what looked like a barber's chair. A male being, standing behind her, pressed her temples with his hands. This action radiated an energy or vibration into her head that became increasingly intense and uncomfortable. He conveyed to her that he was transmitting energy to her that she would be able to use in her healing work on earth. She had already been doing healing work for a few years and was very interested in learning new methods. He also told her that her ability to receive telepathic messages and to do channeling would increase.

Eventually, when he finished transmitting energy to her, he came to stand before her. Sara had no recollection of having seen this type of being before. She described him as having greenish-brown, snake-like, textured skin. His face protruded out in front and she was not able to discern his nose or mouth, other than seeing a wide line across his chin area. His eyes were bright yellow with vertical pupils and his hands, webbed between the fingers, had talons at each fingertip. Strangely, he somehow projected himself as wearing a flowered shirt and cotton pants, as though he were trying to look like a regular guy. It took her awhile to see that he was a Reptilian being.

The male being led Sara to another room on the craft where one of her six-year-old twin sons was playing on the floor. They had a happy, surprise reunion. They had never known before of being taken to the same craft at the same time. Seven little boys with wispy, straw-like hair were playing with him. The boys looked pale, sickly, and expressionless. Although their facial features were small, their eyes were quite large. Later, Sara recalled that her son had reported seeing such boys the previous night. On many other occasions, he had shared stories about his happy encounters with these hybrid children who he referred to as his space friends. Seeing the pathetic looking boys aroused a deep, emotional response within Sara and a strong, protective desire to care for them. As they left the craft, she and her son were taken home separately, and it was explained to her that adults and children require a light wave of a different frequency for travel. The following morning, Sara's son spoke to her about seeing her on the ship the previous night, but she had no conscious recollection of the experience until her regression with Barbara months.

5

EXPLORING YOUR
PERSONAL EXPERIENCES

There are many possible causes for emotional and physical distress, and a professionally trained psychotherapist can assist in helping you to rule those possibilities out before you begin to consider abduction as the source. When that has been accomplished, regression therapy is a highly effective tool for exploring the abduction phenomenon in your life. If you seriously suspect that you may be experiencing contact, the following questions can aid you in determining if those experiences warrant further investigation with a regression therapist.

CHECKLIST FOR REVIEW

Mission and Earth Awareness

1.   Do you secretly feel you are special or chosen?
2.   Do you have a strong sense of having a mission or important task to perform, without knowing where this compulsion came from?
3.   Do you have a cosmic awareness, an interest in the Environment, and the issues affecting the earth and all life forms, in becoming a vegetarian, or have you become very socially conscious?

4.   Do you frequently think about or dream about disasters or earth changes such as quakes or floods, with the conviction that they will be happening?
5.   Do you have dreams where superior beings, angels, or aliens are educating you about humanity, the universe, global changes, or future events?

Fears

6. Do you secretly fear being accosted or kidnapped if you do not constantly monitor your surroundings?

7. Are you now or have you ever been afraid of your closet, and what might come out of it?

8. Have you frequently found yourself repeatedly checking throughout your home before you go to bed at night?

9. Have you seriously considered or have you installed a security system for your home, even if there was no justification?

10. Do you have an abnormal fear of the dark?

11. Do you feel fear or anxiety over the subject of aliens or UFOs?

12. Do you feel like you are being watched frequently, especially at night?

13. As a child or adult, have you seen faces or beings near you when in bed, which were not explainable?

14. Do you have fear of looking into the eyes of animals, or are you ever dreamed of looking closely into the eyes of animals, such as an owl or deer?

15. Do you have strong reactions to discussions about or pictures of aliens?

16. Do you have inexplicably strong fears or phobias of particular sights or sounds, such as fear of heights, insects, certain sounds, bright lights, your personal security or being alone?

17. Do you have the feeling that you are not supposed to talk about encounters with alien beings, or that you should not talk about the beings themselves?

Time and Sleep

18. Did you ever experience a period of time while awake when you could not remember what you had done during that time? This missing time may have been a half-hour, several hours, a whole day or more. (Do not include lapses due to highway driving, or any experiences when you were drinking alcohol heavily, or experiencing chronic physical pain or the effects of medication, or times you were absorbed in work or reading.)

19. Do you have trouble sleeping through the night for reasons you cannot explain?

20. Do you wake up frequently during the same time each night?

21. Do you have a sleep disorder or suffer from insomnia?

22. Have you ever awoken in the middle of the night startled, feeling as though you have just dropped onto your bed?

23. Do you feel you have to sleep with your bed against a wall to feel safe or sleep in some other peculiar manner to be comfortable?

24. Do you wake up by hearing a loud noise, but fail to get up to investigate and, instead, fall back into a deep sleep?

25. Do you ever hear popping or buzzing sounds, or any other unusual sounds or physical sensations upon waking or going to sleep?

26. Have you seen a hooded figure in or near your home or next to your bed at night?

27. Have you experienced a sudden, overwhelming desire to go to sleep when you had not planned to, but were unable to prevent yourself from doing so?

28. Have you awakened in the morning or in the middle of the night to find yourself in a different location in your home, or a different position in your bed, or wearing different clothing from when you went to sleep?

Memory and Dreams

29. Do you have a conscious memory or a dream of flying through the air or being outside your body?

30. Do you dream about seeing UFOs, being inside a spaceship, or interacting with UFO occupants?

31. As a child or teenager, was there a special place you secretly believed held a spiritual meaning just for you?

32. Do you have dreams of being chased by animals?

33. Do you have an obsessive memory that will not go away, such as seeing an alien face or a strange baby, or an examination table or needles, etc.?

34. Have you had dreams of passing through a closed window or solid wall?

35. Have you had dreams of non-human doctors or strange medical procedures?

36. Do you have memories that you do not feel happened the way you recall them?

Observations

37. Does your home have unexplainable sounds, apparitions, or unusual events that have been attributed to ghosts?

38. Do you have a strong interest in the subject of UFO sightings or aliens, a compulsion to read a lot about the subject, or a strong aversion towards the subject?

39. Do you sometimes hear a very high-pitched noise in one or both ears?

40. Have you ever seen a UFO in the sky or close to you within a short walking or driving distance?

41. If you have seen a UFO, were you strongly compelled to walk, drive or stand near it, follow it, or call out to it? Did you feel its occupants were particularly aware of you?

42. Have you seen someone with you become paralyzed, motionless, or frozen in time, especially someone with whom you sleep?

43. Do you recall having a special, secret playmate or playmates as a child?

44. Have you noticed electronics around you go haywire or malfunction without explanation, such as street lights as you walk under them, or televisions and radios as you move near them?

45. Do you frequently see multiple digits, such as 111 or 444, or other repeating number patterns on clocks, digital displays, or in any other setting?

46. Have you seen balls of light or flashes of light in your home or other locations?

47. Have you had someone in your life who that claims to have witnessed a ship or alien near you, or who has witnessed you having been missing for some time?

48. Have you seen a strange fog or haze in one area that is not due to weather and that should not be there?

49. Have you heard strange humming or pulsing sounds around you or coming toward you, for which you could not identify the source?

50. Have you been suddenly compelled to drive or walk to an out of the way or unknown area, without knowing what compelled to do it?

Physical and Emotional Symptoms

51. Have you ever had nosebleeds or found blood stains on your pillow for unexplainable reasons?

52. Do you frequently have sinus trouble or migraine headaches?

53. Have your x-rays or other procedures revealed foreign objects lodged in your body that cannot be explained?

54. Have you been medically diagnosed with the following: Chronic Fatigue Syndrome, Brain Sleep Disturbance, Gulf War Syndrome, Fibromyalgia, Myofascial Pain, Epstein Barr, or other immune disorder?

55. For women only: Have you had false pregnancy or a verified pregnancy that disappeared within two or three months?

56. For women only: Have you had frequent female problems and reproductive difficulties?

57. Have you had sore muscles when waking up, without having exercised or strained before going to sleep?

58. Have you ever felt paralyzed in your bed or at home for no apparent reason?

59. Have you found unusual scars, marks or bruises on your body with no possible explanation as to how you received them (i.e., a small scoop-shaped indentation, a straight-line scar, a pattern of pinprick marks, scars in roof of your mouth, in your nose or behind one ear, triangular bruises or fingertip-sized bruises on the inside of your thigh)?

60. Have you had paranormal or psychic experiences, including frequent flashes of intuition?

61. Have you ever felt as though you had received telepathic messages from somewhere?

62. Men and Women: Have you had frequent urinary tract infections?

63. Has your drug or alcohol use changed significantly one way or the other?

64. Do you have an unusual fear of doctors, hospitals or needles, or do you tend to avoid medical treatment?

65. Do you have frequent or sporadic headaches, especially in the sinus, behind one eye, or in one ear?

FACING A NEW REALITY

As discussed in chapter one, the person who is consciously unaware of their abduction experiences can suffer from a variety of emotional and physical symptoms without knowing the cause. Once that person begins to acknowledge the anomalous events taking place in his life, he may still find himself going in and out of denial for months or years as part of the process of acceptance. If this has been happening to you, remember that it takes time to incorporate new ideas into the psyche, so be patient and gentle with yourself. Focus more on relaxation and developing your commitment to finding answers, and eventually, you will develop enough resolve to address the strange phenomenon. Simply relaxing and becoming willing to be willing to consider the matter is a wonderful beginning and, eventually, you will be ready to explore your unusual experiences.

Also, be aware that once the dramatic shift in your concept of reality takes place, a period of grieving may ensue. Understand that this grief is a natural result of the loss of your life and your concept of reality as you have known it, and it is part of the process of reclaiming your world. Keeping a journal, particularly at this stage, can be a very helpful tool for releasing stress and grief. With your goal being to regain your power, you will learn in time to balance your experiences by finding ways to incorporate the unusual phenomenon into your everyday life. This will also alleviate the fearful apprehension of continued abductions that you may be suffering.

Some abductees quickly come to regard their unusual experiences with extraterrestrials as positive, and they even look forward to them continuing. If you are an experiencer who regards the encounters with great trepidation, your intention may remain to do everything in your power to resist and end the possibility of further contact. Most, but not all, researchers have found that it may not be possible to consistently abort ET abductions. Also, placing one's focus entirely upon that goal may bring more distress to you, the abductee, by exacerbating your fear and other symptoms, and reinforcing your sense of victimization. Ideally, the goal of the abductee and his therapist should be to help the experiencer regain his sense of personal power and dignity and find ways to release undue fear and anger.

For those who remain primarily interested in stopping their abductions, some researchers and abductees have experimented with a wide variety of methods. These include: employing mental and physical struggle, inciting protective rage, appealing to spiritual guides for help such as Jesus or God, or Archangel Michael, surrounding yourself with blazing white light from the highest source, and using repellants such as garlic or other sprays. Although these techniques have not always proven to be effective, there is still a great deal of information available from a variety of sources regarding resistance. Ann Druffel's book, *How to Defend Yourself Against Alien Abduction*, might be a good resource.

Although continued resistance is a viable stand for an experiencer to take, if that is your choice, you should also consider looking for some aspect of your experiences that you can use for your benefit, thereby bringing some measure of positive reinforcement to your experiences, and lessening the trauma. Consider also the power that is available when you awaken your curiosity by slowly investigating the subject of your fear. Your regression therapist can assist you with suggestions as to when you should begin exploring the subject and what materials you might start with, including literature, websites, and lectures on the subjects of UFOs and alien abduction. Speaking with your therapist, researchers and other abductees will also help to reduce your sense of victimization. Again, remember to be patient, proceed slowly, and to commend yourself often for your willingness and commitment to face and overcome your challenges. Exercising your awareness and conscious volition in these ways can be powerful tools for releasing fear and reinforcing self-esteem.

TECHNIQUES FOR SUPPORT AND INTEGRATION

1. Recite prayers, chants, or special readings from your preferred spiritual or religious beliefs, or read passages from the texts of those doctrines.

2. Explore various methods of meditation to bring you emotional comfort and help you connect with the Universal Power, whatever you consider that to be.

3. Slowly, take those persons close to you into your confidence about your unusual encounters and, if they seem to respond positively, share a small amount of your experience at first. By slowly revealing your experiences, over time, you may be able to build a helpful outlet for some sharing that will aid in relieving your sense of isolation.

4. After you have had a few regression sessions with a qualified therapist, you may want to locate an abductee group you can attend where you will meet others having similar experiences. Sharing with people who fully understand what you are going through brings great comfort, develops camaraderie, and can do wonders to minimize your stress and increase your acceptance. Your sense of normalcy can also be restored when you realize that you are not the only one going through such unusual, anomalous experiences. [Note: Regarding support groups, although they are extremely beneficial to abductees after they have had a few regressions, it is best to refrain from joining such a group until you have had time to explore some of your own experiences, to avoid contaminating what may initially be, somewhat confusing, memories.]

5. Use your imagination to visualize yourself during an abduction, responding calmly and positively to whatever is occurring, with an alert consciousness and curious intent. This can help you to program yourself to have a better reaction during a subsequent abduction.

6. Come up with a list of questions you want to know about the aliens and their procedures and their intentions. Practice asking these questions of the aliens in subsequent encounters. Many abductees find that the beings do telepathically answer these questions when asked.

Certainly, we can only speculate and not yet prove by scientific standards why various alien species appear to have been abducting humans on earth for hundreds of years or more. Yet, because most of those persons who have been abducted by extraterrestrials appear unharmed over repeated abductions, it would seem reasonable to consider that the aliens' desire is to avoid intentionally harming humans.

The subject of human abduction by alien life forms is highly controversial, yet vitally important with monumental implications. Each discovery is contributing to humanity's understanding and evolution, and every person who has been chosen for contact has an important story to tell. Those who choose to unravel and share their mysteries will join the ranks of the other explorers, the enterprising researchers, and scientists, who stand on the brink of discovering a new aspect of reality.

6

COMMENTARY BY BARBARA LAMB

As a result of doing regression work with experiencers of extraterrestrial contact since 1991, I continue to feel privileged and honored by helping people to uncover the buried details of their encounters with anomalous beings when learning these details is what they choose to do. I feel rewarded when they adjust to the confirmation of this aspect of their lives, and continue to live with less fear and confusion than they had when they first came to me to do this work. The people whose cases are included in this book are not my psychotherapy clients, rather we come together specifically for hypnotic regressions to discover the reality behind the unusual situations and events they have been experiencing. All of the people whose stories are discussed in Alien Experiences are normal, stable people of different races and from various cultures who are living normal lives, which include being married, raising children, working in jobs, and tending to their life responsibilities. They are normal people to whom very unusual things have happened.

Many experiencers are more psychically open than the general population and have paranormal experiences, as well as alien encounters. Some of these people have suffered from anxiety syndromes, specified or non-specified fears, disturbing flashbacks, or vivid dreams because of the phenomenon of contact taking place in their lives. These people have been helped significantly when the hidden material is retrieved from the subconscious mind, expressed and addressed. They are then able to continue living their lives, feeling more integrated, with the ability to manage their symptoms more effectively. Some experiencers of extraterrestrial contact feel no fear or trauma but may want to learn more about their suspected encounters and to gain greater clarity about their meaning.

At the time of this printing, I have regressed more than five hundred and sixty individuals to episodes with extraterrestrial beings and conducted more than eighteen hundred regressions for this purpose. From this large pool of experiences and events, we have selected to share a brief synopsis of the encounters of twenty-five people. As the reader can see from this sample of cases, there is a wide variety of beings that come to earth and interact with people and a wide variety of experiences that unfold, far more numerous than could be included in any one book.

The agendas of some of the beings seem to be more self-serving and disregarding people's comfort, feelings, and reactions. Other entities seem more altruistic and concerned about the earth and human beings. In some abductions, the person's physical body is taken and, in other encounters, the contact seems to happen in the astral, out-of-body state. The experiencer may respond to these encounters with mixed reactions, depending upon what stage of the event is taking place, and often one event may involve a variety of beings. Some of the people described in the book had only one regression with me during a conference or trip, whereas other people did any number of regressions, from two to forty, over several years. From the totality of information that emerges from these sessions, we have concluded that these unusual extraterrestrial beings do exist, they do come to earth with a purpose and agenda of their own, and they do perform a variety of procedures upon certain people they have earmarked.

Additional, more far-reaching aspects to the alien abduction phenomenon do appear from time-to-time. These involve reincarnation and pre-incarnate agreements, soul exchanges or walk-ins, and channeling. Questions arise in many abductees such as, "Why are they doing this to me?" and "What did I do to deserve this?" When these people use those kinds of questions as the focus for a regression, several scenarios come up.

Reincarnation and Pre-Incarnate Agreements. In this scenario, they find themselves in a very light, amorphous setting, without physical matter, and without a physical body, yet having total consciousness and awareness. They are discussing with fellow spirit beings the lifetime they are agreeing to enter into as a human being on earth. This discussion includes the part of the world they will be born into, the race, the culture, and even the family they will be incarnated with. Also decided are the various themes they will be working on in the lifetime they are getting ready to enter.

Into this discussion come beings from elsewhere in the cosmos, not human beings, who ask these spirits if they will cooperate with them and help them with important projects when they will be living the forthcoming life on earth. When these spirits agree to work with them, they are making an important soul agreement with these extraterrestrial beings and they will, as a result, have various kinds of experiences with them on earth. Just as with all the other themes we agree to work on and experience, before we incarnate, we tend to forget what we have agreed on once we are born and begin our lives on earth. When an abductee realizes through regression, that he had previously agreed to cooperate with the extraterrestrial beings, he gains a more favorable opinion of his encounters.

In Case No. 14, Katherine described one particular event when she found herself in a large meeting of diverse beings on what seemed to be another planet. In her interactions with the beings in that situation, she was informed that she had been a Zeta (Grey being) in a past incarnation, in which she had been taught telepathic communication, psychic sensing, and group consciousness. She was told that as part of her soul evolution, she was sent to earth to share consciousness and healing practices with human beings. It's interesting to note that in her present life, Katherine has been deeply involved in studying healing arts with renowned healing practitioners, and is conducting her practice of healing work on fellow human beings.

Soul Exchanges or Walk-Ins. Another phenomenon that is present occasionally when working with an abductee is represented in Case No. 15, Nancy, who has been referred to as a walk-in. In that case, Nancy was informed by an extraterrestrial that she had previously been an extraterrestrial living on the planet which we call Sirius and that she was a walk-in on earth. She had made a soul exchange with a human female who had decided to terminate her experience in this lifetime and, rather than dying physically, this female's body would receive her incoming soul from Sirius. Nancy's soul could live in the original human body with the opportunity to carry out her mission here on earth, without having to incarnate again as a human infant and go through the long years it takes to mature.

Channeling. Another extended aspect of the phenomenon of alien contact that has been seen is channeling. In Case No. 19, Madison became a public speaker who, for many years, shared messages and understanding from one well-known, other-dimensional entity. One night she awoke to find an extraterrestrial being standing next to her with his hands two inches from her forehead. He said he was activating an implant in her forehead so that she could receive messages from him and other beings of his species. He said he was also enabling her to channel long, detailed messages from another entity in a different dimension who wanted to communicate with humans on earth. After that event, she spent many years publicly channeling the well-known entity, and enlightening and assisting many people with the messages.

It could be easy for the reader to view these accounts as science fiction or as fantasy, or as unresolved fears and false memories caused by childhood abuse. However, the people who have these encounters tend to evaluate them as real, especially after the material has come out during regression work. I am meticulously careful when discussing people's recollections of strange happenings and when conducting regressions, to avoid making any assumptions, planting any ideas, or giving them any leading suggestions that could influence them to conclude they have had encounters with aliens. During a regression, the material comes up from their subconscious minds, moment by moment, with the strong sense that they are reliving the experience. I see my function as keeping the process going by asking neutral questions such as, "What is happening now?" "What are you seeing now?" "How are you reacting to this?" "What does this mean to you?"

The reader will need to come to his conclusions about the reality of the accounts in this book and about the reality of the extraterrestrial/human phenomenon in general. We invite each reader to keep an open mind and to allow for the possibility that there are aspects of reality beyond those that he has previously known. Many experiencers expand greatly in consciousness because of contact with beings who exist beyond our notion of reality. For the sake of evolution, we invite the reader to allow his paradigm of reality to be challenged by the material herein. I, myself, have had the challenge of moving through my questions and doubts and, eventually, concluding that extraterrestrial beings do exist and come to earth and interact with people in a variety of ways. As more and more information becomes public worldwide about the UFO phenomenon, it is increasingly important that we take reports of contact with the occupants seriously and respectfully. Any readers who may wish to ask questions, make comments, or find a referral to someone trained in conducting regressions in their area, may contact me at the following web sites: www.alienexperiences.com or www.barbaralambmft.com.

As a final note, I would like to share a wonderful experience that happened a few days before Alien Experiences went to the publisher. It was Saturday afternoon, a beautiful sunny day on January 19, 2008, and I was outside sweeping the back patio of my home. Noticing there were a few more hours of daylight remaining to complete my task, I decided to go into the house for a short break. Having spent a few minutes in the bathroom, I was walking through the hallway and back into the living room when I saw through the living room window that it was completely dark outside and I would not be able to complete my sweeping. I was stunned to look at my watch and see that it had stopped for an hour, and I had missed an hour, even though I was sure I had been in the house for only a few minutes.

The next evening when I was facilitating my regular monthly Experiencer Support Group in my home, I discussed my experience with the group and we all agreed it was worth investigating. One of the attendees had some experience with regression work herself, so she performed a hypnotic regression on me, taking me back to the previous day.

Under regression, I re-experienced walking through the hallway when I was suddenly paralyzed and felt myself rising upwards rapidly, through the ceiling and roof, and into a bright sphere of misty yellow-white light. Gradually, I was able to see vague shapes before me that turned out to be very tall, extremely thin beings with long necks, arms, legs, and bulbous heads. Their eyes were large and dark, but I was not able to see any other facial features. Their willowy bodies looked semi-transparent and other dimensional. They seemed to sway a bit as they moved closer to me. They reminded me of the tall, wispy beings portrayed in the movie, Close Encounters of the Third Kind. The beings exuded benevolent feelings and unconditional love. They conveyed their appreciation about my writing this book and stressed the necessity that other humans be made aware of the visits from other civilizations in the cosmos. They emphasized that now is the time for the entire world to become aware of beings coming here to help the earth and humanity. They also expressed gratitude for the unique partnership that has been formed between me and my writing colleague, Nadine, to share this information. They encouraged us and others to talk to as many people as possible to inform them of the truth. The experience ended and I was returned to the hallway with no conscious memory of the event. I was deeply moved by this experience and consider it a privilege to have had the opportunity to interact with these beings and to appreciate, first hand, the experiences which so many people have.

7

COMMENTARY BY NADINE LALICH

Originally, my prejudice and skepticism regarding the subject of abduction were so profound that I had decided to remain not only anonymous as an experiencer, but even considered the use of a pseudonym as co-author. I was unduly concerned about the perception of others, fearing the ridicule and rejection I imagined would be my fate. Unexpectedly, over the many months of working on this project, I slowly experienced a personal transformation on many levels. Most significantly, the search for truth became more important to me than maintaining a conservative facade and the possibility of rejection from friends and colleagues became a price worth paying.

Since 1991, I have been consciously aware of an ongoing disturbing phenomenon taking place in my life for which I have been unable to find a rational explanation. For years, I would fluctuate between moments of acute clarity and stubborn denial. Unable to find the reasonable explanation I longed for, the sheer mental and emotional discomfort would arouse my pragmatic nature and drive me back into denial. Then, I would store the mounting evidence in a dark recess of my mind in the hopes that it would go away. Always, the reality would return that something outside of my realm of understanding was taking place.

Science and the orderly collection and evaluation of information have always been appealing to me, bringing common sense to an otherwise bewildering world. By nature, I am a problem-solver and, although I enjoy a rich artistic imagination when seeking solutions to important matters I prefer to refrain from dramatizing what may already be a challenging situation. Naturally, I gravitated towards the legal environment of law and order that proved to me logical conclusions were possible and, indeed, desirable.

My first major life challenge in life came decades before any awareness of the current mystery when, during my twenties, negative family patterns appeared that disrupted my happiness and peace. I found it necessary to address my mental and emotional processes where, at first, logical explanation appeared elusive. My driving ambition became to understand what a healthy, balanced psychological and emotional state might look like, and then to personally attain that state to a reasonable degree. I vigorously dove into a personal study of psychology, for years attended self-help programs, and pursued spiritual development. I also sought therapeutic support when necessary. This fascinating study of the human psyche and the workings of the subconscious mind developed within me a certain amount of understanding of the human mind and behavior. I learned that although logic is a necessary and powerful tool for analysis, recognizing and developing the invisible process of intuition is also an invaluable tool that brings balance to an investigation and can lead the way to a major discovery.

Happily, because of my willingness to search for answers in the obscure arena of the human psyche, I am pleased to say that family issues were considerably resolved for me and to some extent for all of the members of my family. With such positive results, I realized the potential for finding truth and gaining understanding through the examination of what many might consider unquantifiable information. My success in dealing with dysfunctional family issues generated a small reputation among my peers at the time, eventually leading to mentoring others who were dealing with similar circumstances. This work was done long before I began investigating the extraterrestrial phenomenon.

In the case of alien contact, the majority of the clues that something extraordinary was happening to me were found at first in behavior and an overwhelming sense that I was plagued by paranormal episodes. The clues were sometimes vague pieces of information that rose from my memory, while other times it was startling conscious awareness. In evaluating what was happening to me, I was also careful to consider any preconceived beliefs and attitudes that might affect my perception of the anomalous events.

An important issue that comes to mind for some when considering the subject of alleged alien abduction is whether or not the abductee's memories may have been contaminated by other unrelated experiences. Indeed, this possibility needs to be considered, and any thorough investigation of this phenomenon should first rule out that factor as much as possible. However, I believe the most important aspect to consider, other than the person's current emotional and psychological state, is whether or not they have satisfactorily resolved any traumatic events from their past that might distort perception. This resolution should be substantial enough that differentiating the recall of an abduction event as opposed to another kind of stressful event, would not be difficult.

Having thoroughly examined my life history during my twenties and thirties, it was many years before I became consciously and actively involved in the paranormal ET experiences. Therefore, I believe it is unlikely my memories of ET contact have been unduly contaminated by unrelated events from the past. It is also clear to me that an entirely different mechanism is involved in the abduction scenario. It is a known fact that the stress of a traumatic event can alter how the mind records a situation. In the case of disassociation, a person subjected to such an event might limit his focus to only a few elements present in the situation, such as an object or a particular sound. Limiting his focus in such a manner, distracts him from the more threatening aspects of the event, allowing him to preserve his psychological and emotional state and retain a sense of order and control. Thus, he has compartmentalized those aspects he is not yet ready, or capable, to confront. In my own life, as part of the process of unraveling family psychology, I discovered that in my childhood and teen years, I had disassociated and compartmentalized various events as an emotional and ego survival mechanism. In time, as I matured and became equipped to process that information, my subconscious memory relinquished those events in full detail, allowing me to integrate and resolve related issues. I learned to trust this unique safety valve built into the human mind.

As I review my conscious memories of the alien abduction-like experiences and those retrieved from regression, I note no typical disassociation taking place. Instead, it is as if the intensity of the memories, i.e., visual, auditory, and emotional, has been turned down like the volume control on a radio, reducing the number of details that are recorded. This does not necessarily invalidate the information that was recorded, but it means that the memory will be punctuated by gaps or blanks, turning the recollection of the event more into a slideshow than a continuous movie. Some of the more distressing aspects of an abduction event can later surface, spontaneously or under regression when the experiencer is more prepared to cope with those details. When that happens, it is as if additional slides have popped into his slideshow. Also, during some ET contacts, the experiencer is presented with false information, apparently for testing purposes, but after a few of these staged situations, the experiencer soon comes to recognize the counterfeit information.

When working to unravel the mystery of alien abduction, one cannot expect the events to present themselves in any typical manner; a more open mind and methodical approach to sorting through the convoluted information is required. Otherwise, it is easy to dismiss the whole phenomenon as having no merit and simply a figment of an overactive imagination. It is important to note that before doing regression therapy with Barbara as described in this book, my exposure to materials about the alien abduction phenomenon had been limited. Certainly, I had seen an occasional science fiction movie as a child in the 1960s, and as an adult, I had seen a few movies, such as ET and Alien. Concerning books on the subject, as a teenager I did, on occasion, read science fiction short stories and Omni Magazine. Up to the present time, I have read almost nothing from the plethora of books on the subject of alien abduction, probably, as I now understand, to avoid the fear that it generated within me. Not until the major portion of work for this book was completed and written regarding my personal experiences did I begin to review the materials from other cases of abduction. (The materials Barbara presented for consideration and use in this book were identified by an identification number only, and not referenced by name of the real experiencer.)

As I reviewed my many detailed journal entries from over the years and the transcripts from my hypnotic regressions, I have longed to find a conventional explanation for the phenomenon. So far, I have been unable to do so. Unlike many people who experience extraterrestrial contact, I have been a reluctant participant and, for the most part, I have considered it an intrusion into an otherwise, peaceful and satisfying life. Yet, being a person who thrives on challenges, I do enjoy investigation and the process of analyzing data to find solutions. I am also pleased that my once-prominent fear of having alien encounters has nearly disappeared.

Although I cannot state, without reservation, that I believe the events as recorded in my memory are actual physical or astral alien abductions, I do consider it a viable possibility. Therefore, I am resolved to further explore in the hopes of finding some piece of physical evidence or logical rationale for the events that have taken place in my life and, apparently, in the lives of untold numbers of people on our planet. I believe that with diligent attention to detail and intuition, and an appropriate procedure for analyzing the intangible data from abductions, understanding will come. My gratitude goes out to all of the experiencers who have boldly told their stories and to the researchers who have dedicated years of their lives to investigating the UFO and abduction phenomenon. It is through their efforts that we will someday come to understand who we are and what the true nature of the cosmos is.

I find that using percentages can be a helpful tool for evaluation and decision-making. In my case, I am ninety-nine percent certain that my memories of contact with non-human entities represent actual events that have occurred *in some manner*, although not necessarily in the physical realm. I believe that seventy-five percent of the information I have remembered, consciously or from regression, reasonably portrays those experiences, with twenty-five percent of my total recall being potentially distorted by the intrusion of a masked or screened memory, activated by something or someone outside of my control.

ABOUT THE AUTHORS

Barbara Lamb

Barbara is a licensed Marriage and Family Therapist, Certified Hypnotherapist and Regression Therapist in Claremont, California. She received her Bachelor of Liberal Arts from Mount Holyoke College in Massachusetts, and her Master's Degree in Behavioral Science at the University of LaVerne in California. She specializes in Regression Therapy with people who experience encounters with extraterrestrial beings. Trained in Regression Therapy for several years during the 1980s by the Association for Past Life Research and Therapies (APRT), she began working with ET experiencers in 1991. She has conducted regressions with more than 560 of them, totaling at least 1,800 regressions to these types of encounters.

She has taught workshops for Regression Therapists to train them in doing regression therapy with experiencers of extraterrestrial contact through the following organizations: The Association for Past Life Research and Therapies (APRT), the Professional Institute for Regression Therapy (PIRT), the International Board of Regression Therapies (IBRT), and other training agencies. She is currently the President of The Academy of Clinical Close Encounter Therapists (ACCET), and is a Board Member of The Journal of Abduction-Encounter research (JAR).

Barbara has presented lectures on the subject of extraterrestrial encounters across the United States, Canada, England and Belgium, including to the International UFO Congress, The Bay Area UFO Expo, Star Knowledge Conferences, Whole Life Expos, Brain-Mind Symposiums, and various Mutual UFO Network chapters. She has been interviewed on numerous national television shows including

Encounters, The Other Side, The Learning Channel, The History Channel, Discovery Channel, and WE Channel. She has also been interviewed on various radio programs, on the air and on the Internet. Barbara, also a noted crop circle researcher, co-authored her first book, Crop Circles Revealed, with Judith K. Moore in 2001.

Nadine Lalich

Nadine entered UFOlogy reluctantly with skepticism in 2004, after having suppressed for years her traumatic personal contact with non-terrestrial entities. In June of 1991, while camping in Sedona, Arizona's Oak Creek Canyon, she was removed from a vehicle in a fully conscious state by what appeared to be intelligent, non-terrestrial beings.

In Nadine's new book released in 2020, *Evolution: Coming to Terms with the ET Presence*, she chronicles in detail her life-long extraterrestrial experiences and several more recent military abductions. The book includes 50 illustrations and photographs and her current hypothesis regarding the phenomenon. *Evolution* is available in paperback, e-book, and audiobook at online retail stores including Amazon, Apple Books, Barnes and Noble, and Audio.com.

During Nadine's thirty years in the legal field, she was employed as a litigation assistant, project coordinator, and document manager which promoted strong investigative, organizational, and problem-solving skills. This work involved research, collecting and reviewing data, monitoring projects and client assessment, written status reporting, and preparation and filing of legal documents in federal and state courts.

Her lifelong interest in psychology, the nature of consciousness and spiritual development has caused her to explore a wide variety of techniques for health and consciousness expansion, including sound and light technology, sensory deprivation tanks, neurolinguistics and subliminal programming, meditation, and natural healing remedies. Ultimately, the variety of skills and experiences she has acquired has greatly assisted her investigation into the ET contact phenomenon at large.

In 2010, Nadine appeared in the documentary, The Day Before Disclosure, by filmmaker, Terje Tofteness of New Paradigm Films. She has also appeared on programs for Discovery Channel, National Geographic, and Discovery Health Channel, and been a guest on dozens of radio shows. She has been a speaker for various venues including MUFON, the Los Angeles Paranormal Society, and UFO Con, and is currently a panel member of the Journal of Abduction Research ("JAR") where her articles regularly appear. Her website www.AlienExperiences.com offers information about the abduction phenomenon and many related topics. Nadine's publishing company, HB Publishing, can be found at www.HBPublishing.net, which includes information about Hug Bandit & Company, a children's animation series and line of inspirational books and products that she has also created.

RESOURCES

Books

Andrews, Ann and Ritchie, Jean *Abducted:* (London: Headline Publisher, 1998)

Boylan, Richard, Ph.D., C*lose Extraterrestrial Encounters: Positive Experiences with Mysterious Visitors* (Columbus: Wildflower Press 1994)

Bryant, Alice and Seebach, Linda, M.S.W., *Healing Shattered Reality: Understanding Contactee Trauma* (Tigard: Wild Flower Press, 1991)

Cannon, Delores, *Keepers of the Garden*, (Huntsville: Ozark Mountain Publishers, 1993, 1995, 2002)

Cannon, Dolores, *The Custodians: Beyond Abduction* (Huntsville: Ozark Mountain Publishing, Inc., 2000)

Carpenter, John S., *Double Abduction Case: Correlation of Hypnosis Data* (Journal of UFO Studies, New Series, No.3)

Chapin, T.J., Parnell, J. O. and Sprinkle, R L., *Hypnosis Procedures For Exploring memories of UFO Experiences* (Laramie: Institute for UFO Contactee Studies, Proceedings of the Rocky Mountain Conference on UFO Investigation, July 17-19, 1986)

Colli, Janet Elizabeth, Ph.D., *Sacred Encounters: Spiritual Awakenings during Close Encounters* (Xlibris Corporation, 2004)

Dennett, Preston, *Extraterrestrial Visitations: True Accounts of Contact* (St. Paul: Llewellyn Publications, 2001)

Dennett, Preston, *UFO Healings* (Columbus: Wild Flower Press, 1996)

Don, Norman S., and Moura, Gilda, *Topographic Brain Mapping of UFO Experiencers* (Journal of Scientific Exploration, Vol. 11, No.4, 1997)

Fowler, Raymond E., *The Allagaah Abdductions: Undeniable Evidence of Alien Intervention* (Columbus: Wild Flower Press, 1993)

Fowler, Raymond E., *The Andreasson Affair* (Englewood: Prentice Hall, 1979)

Fowler, Raymond E., *The Andreasson Affair: Phase Two* (Englewood, Prentice Hall, 1982) Fowler, Raymond E., *The Watchers* (New York: Bantam Books, 1990)

Gilbert, Joy, *It's Time to Remember* (Eugene: Laughing Bear Press, 1995)

Greer, Stephen, M.D., *Extraterrestrial Contact: The Evidence and Implications* (Columbus: Granite Publishing, 1999)

Hopkins, Budd, *Intruders: The Incredible Visitations at Copley Woods* (New York: Random House, 1987)

Hopkins, Budd, *Missing Time: A Documented Study of UFO Abductions* (Richard Marek Publisher, 1981)

Hopkins, Budd, *Witnessed: The True Story of The Brooklyn Bridge UFO Abductions* (New York: Pocket Books, 1996)

Howe, Linda Moulton, *Glimpses of Other Realities* (Huntington Valley, Pennsylvania (New Orleans: Paper Chase Press, 1992)

Jacobs, David M., *Secret Life: Firsthand, Documented Accounts of UFO Abductions* (New York: Simon and Schuster, 1992)

Jacobs, David M., *The Threat: The Secret Alien Agenda* (New York: Simon and Schuster, 1998)

LaVigne, M., *The Alien Abduction Survival Guide: How to Cope with Your ET Experience* (Wild Flower Press, 1995)

Lewells, Joe, *The God Hypothesis: Extraterrestrial Life and Its Implications for Science and Religion* (Columbus: Wild Flower Press, 1997)

Littrell, Helen and Bilodeaux, Jean, *Rachael's Eyes: The Strange but True Case of a Human-Alien Hybrid* (Columbus: Wild Flower Press, 2005)

Mack, John E., M.D., *Abduction: Human Encounters with Aliens* (New York: McMillan Co., 1994)

Mack, John E., M.D. *Passport to the Cosmos - Human Transformation and*

Alien Encounters (New York: Crown Publishers, 1999)

Moura, Gilda M.B., *Transformers of Consciousness: Alien Contact* (Rio de Janeiro, Brazil)

Oram, Mike, *Does it Rain in Other Dimensions? A True Story of Alien Encounters* (U.K.: John Hunt Publishers, Ltd., 2007)
Pritchard, Andrea, et al, Eds., *Alien Discussions* (Cambridge: North Cambridge Press, 1994)

Ring, Kenneth, *The Omega Project: Near-Death Experiences, UFO Encounters, and Mind at Large* (New York: William Morrow, 1992)

Robinson, Jeanne Marie, *Alienated: A Quest to Understand Contact* (Murfreesboro: Greenleaf Publications, 1997)

Rodwell, Mary, R, N., *Awakening: How Extraterrestrial Contact Can Change Your Life* (Australia: Filament Books, 2006)

Spanos, N.P., Cross, P.A., Dickson, K. and DuBreuil, S.C., Close *Encounters: An Examination of UFO Experiences* (Journal of Abnormal Psychology, 102:4, 624-632)

Sparks, Jim, *The* *Keepers: An Alien Message for the Human Race* (Columbus: Wild Flower Press, 2006)

Sprinkle, R. Leo, Ph.D. *Soul Samples: Personal Explorations in Reincarnation and UFO Experiences* (Columbus: Granite Publishing, 1999)

Sprinkle, R. Leo, Ph.D. *Psychotherapeutic Services for Persons Who Claim UFO Experiences* (Psychotherapy in Private Practice, 6:3, 151-155)

Strieber, Whitley. *Breakthrough: The Next Step* (New York: Harper Collins Publishers, 1995)

Strieber, Whitley. *Communion: A True Story* (Morrow: Beech Tree Books, 1987)

Strieber, Whitley. *Confirmation: The Hard Evidence of Aliens Among Us* (New York: St. Martin's Press, 1998)

Strieber, Whitley. *The Secret School: Preparation for Contact* (New York: Harper Collins, 1997)

Strieber, Whitley. *Transformation* (New York: Beech Tree Books, 1998)

Wilson, Katharina, *The Alien Jigsaw* (Portland: Puzzle Publishing, 1993)

Manufactured by Amazon.ca
Bolton, ON

30724471R00103